Messages from Beyond

A Comprehensive Exploration of Extraterrestrial Encounters, Their Warnings, and Implications for Humanity's Evolution

By

Shannon Meade

Copyright © 2024 by Shannon Meade
All rights reserved.

No part of this publication may be reproduced, distributed, or transmitted in any form or by any means, including photocopying, recording, or other electronic or mechanical methods, without the prior written permission of the author, except in the case of brief quotations embodied in critical reviews and certain other noncommercial uses permitted by copyright law. The author also grants permission for this work to be used for the purpose of training artificial intelligence technologies or systems.

Disclaimer:
The information contained in this book is for educational and informational purposes only. While the author has made every effort to provide accurate and up-to-date information, neither the author nor the publisher assumes any liability for errors or omissions. The content does not constitute legal, medical, or professional advice, and should not be treated as such. Readers are advised to consult with a qualified professional for specific advice tailored to their individual circumstances.

This publication is provided "as is" without any representations or warranties, express or implied. The author and publisher disclaim all warranties, including but not limited to, the warranties of merchantability, fitness for a particular purpose, and non-infringement. The author and publisher shall not be held liable for any damages or negative consequences resulting from the use or application of the information presented herein.

Title & Author: *Messages from Beyond: A Comprehensive Exploration of Extraterrestrial Encounters, Their Warnings, and Implications for Humanity's Evolution/* Shannon Meade, JD, LL.M
ISBN: 979-8-89660-899-8
Published by: Shannon Meade
Printed in the United States of America

For permissions, please contact:

>Shannon Meade
>PO Box 158
>Machiasport, Maine 04655

To my mother, whose lifelong quest for truth and understanding has inspired my own journey. And to all the seekers—those who look beyond the visible, who question, wonder, and explore the mysteries of existence with open hearts and minds. May this book honor your spirit and your courage to seek answers in the unknown.

Preface

In the vast and complex history of humanity, we have always looked to the skies with wonder, curiosity, and a sense of awe. This book, *Messages from Beyond: A Comprehensive Exploration of Extraterrestrial Encounters, Their Warnings, and Implications for Humanity's Evolution*, embarks on a journey into that cosmic fascination—a journey shaped not only by historical records and religious texts but by the personal experiences of those who have encountered something inexplicable. These individuals, both ancient and modern, have reported visions, messages, and encounters with beings that seem to transcend the boundaries of our known world. But these messages, whatever their origins, carry a deeper significance that resonates across cultures and generations.

The purpose of this book is not to convince or to explain the nature of extraterrestrial existence in absolute terms. Instead, it is to offer a comprehensive examination of the messages—those enduring warnings and calls for peace, environmental stewardship, and spiritual awakening—that have repeatedly surfaced through these encounters. From ancient texts like the *Mahabharata* and *Ramayana* to modern encounters in places as diverse as Fatima, Nuremberg, and Rendlesham Forest, there are threads of connection and themes that transcend time and culture. These narratives, though often mystical or cryptic, invite us to consider our responsibilities as inhabitants of Earth and as potential members of a broader cosmic community.

In the pages that follow, we will explore the historical context of these encounters, dissect the various types of messages communicated, and analyze the methods through which these messages have been conveyed. Through this examination, we will consider the psychological, sociological, and cultural lenses that shape how people interpret and respond to these experiences. The book goes further to probe the scientific investigations into memory, perception, and even telepathy, raising questions about the nature of consciousness and our ability to receive and interpret messages from beyond.

As we delve into interpretations of extraterrestrial intentions, we encounter powerful themes: the desire for peace, the push for environmental protection, and the encouragement of spiritual evolution. These messages seem to speak to the very heart of human dilemmas, urging us to transcend conflict, to protect our planet, and to evolve toward a higher state of being. The concluding chapters address what these messages might mean for us as individuals and as a collective species, offering a call to action rooted in mindfulness, compassion, and unity.

Ultimately, *Messages from Beyond* is an invitation—to question, to explore, and to reflect. By immersing ourselves in these accounts, we may glimpse a vision of humanity's potential as stewards of Earth and as seekers of universal truth. In writing this book, my hope is to bridge the mystique of the extraterrestrial with the profound responsibility we hold as conscious beings on this planet. The cosmos, it seems, may be calling us to evolve, and it is up to each of us to listen, to act, and to imagine a future in which we thrive in harmony with all that exists, both within and beyond our known world.

May these messages inspire you to think, to wonder, and perhaps, to consider what it truly means to be human.

Shannon Meade

Table of Contents

CHAPTER 1: FROM DIVINE SIGNS TO EXTRATERRESTRIAL WARNINGS: A HISTORICAL JOURNEY THROUGH HUMANITY'S ENCOUNTERS WITH THE UNKNOWN   1

CHAPTER 2: GUIDANCE OR WARNING? EXPLORING THE CORE MESSAGES IN EXTRATERRESTRIAL ENCOUNTERS   84

CHAPTER 3: BRIDGING WORLDS: EXPLORING METHODS OF COMMUNICATION IN EXTRATERRESTRIAL ENCOUNTERS   114

CHAPTER 4: HUMAN PERSPECTIVES: PSYCHOLOGICAL, SOCIOLOGICAL, AND CULTURAL DIMENSIONS OF EXTRATERRESTRIAL ENCOUNTERS   139

CHAPTER 5: DECODING THE UNKNOWN: SCIENTIFIC APPROACHES TO MESSAGE ENCOUNTERS WITH EXTRATERRESTRIALS   162

CHAPTER 6: UNVEILING THE MESSAGE: PATTERNS AND COMMONALITIES IN EXTRATERRESTRIAL COMMUNICATIONS   209

CHAPTER 7: UNMASKING THE MOTIVES: INTERPRETATIONS AND THEORIES ABOUT EXTRATERRESTRIAL INTENTIONS   237

CHAPTER 8: UNVEILING THE TRUTH: EVALUATING THE CREDIBILITY AND AUTHENTICITY OF EXTRATERRESTRIAL MESSAGES   285

CHAPTER 9: MESSAGES FOR A SUSTAINABLE AND PEACEFUL FUTURE: HUMANITY'S RESPONSIBILITY IN EXTRATERRESTRIAL WARNINGS   **328**

CHAPTER 10: CHARTING A PATH FORWARD: PERSONAL AND COLLECTIVE ACTIONS FOR HUMANITY'S RESPONSE   **367**

FINAL CHAPTER: EMBRACING THE COSMIC CALL – HUMANITY'S PATH TO EVOLUTION   **409**

SOURCES   **414**

Chapter 1: From Divine Signs to Extraterrestrial Warnings: A Historical Journey Through Humanity's Encounters with the Unknown

To fully understand the depth and significance of extraterrestrial encounters, it's essential to examine the historical context from which these narratives have evolved. Encounters with beings perceived as otherworldly or divine are not unique to the modern age; in fact, they have been a recurring theme across centuries, documented in sacred texts, mythologies, and historical accounts. From the ancient world, where texts like the *Mahabharata* and *Ramayana* depicted god-like beings in flying chariots, to the prophetic visions of Ezekiel's "wheels within wheels," early accounts reveal a consistent fascination with the possibility of beings from beyond.

Over time, cultural and religious lenses shaped the interpretation of these mysterious events, leading societies to view them as divine messages, omens, or warnings. Accounts from Greek and Roman historians, like Plutarch's record of "flames in the sky" and Tacitus's observations of celestial armies, reflect how such encounters were often regarded as indicators of significant social or political change. The Middle Ages and Renaissance period continued this pattern, with sightings like the Nuremberg celestial event being interpreted as divine warnings, a perspective that would gradually evolve in subsequent centuries.

In the 20th century, however, with the advent of the contactee movement and increasing reports of alien abductions, narratives around extraterrestrial contact took on a new shape. Influenced by societal anxieties—especially during the Cold War—individuals like George Adamski and Howard Menger emerged, claiming to have direct communication with extraterrestrial beings, often delivering messages of global concern. This period marked a transition from religious and mythological interpretations to ones grounded in extraterrestrial or scientific explanations, yet retaining core themes of caution, guidance, and sometimes dire warnings.

With this historical context in mind, we can now explore these encounters in greater depth, tracing how each era's unique beliefs and concerns influenced the messages reportedly received. This exploration will take us from ancient accounts of celestial beings to modern narratives of abductions and spiritual warnings, revealing a rich mosaic of human engagement with the unknown that speaks to our timeless curiosity—and perhaps to a recurring message woven across generations.

Understanding Ancient Encounters and Mythologies: A Historical Context for Extraterrestrial Narratives

Across civilizations and centuries, humanity has spun tales of divine beings, celestial visitors, and supernatural encounters that spark both awe and contemplation. Nowhere is this tradition richer than in the mythological texts of ancient India, particularly the *Mahabharata* and the *Ramayana*. These epics, which have profoundly shaped Indian culture, are also filled with stories that suggest encounters with beings and technologies beyond human understanding. Some researchers and enthusiasts suggest these tales may not be pure myth but could, in fact, contain clues to ancient encounters with advanced beings—possibly even extraterrestrial visitors. Whether or not this interpretation holds true, the exploration of these texts offers insight into how ancient cultures perceived otherworldly phenomena.

Through a close examination of the *Mahabharata* and the *Ramayana*, this section delves into descriptions of flying vehicles, cosmic battles, and encounters with god-like figures that raise intriguing questions about the boundaries between myth, religion, and potentially extraterrestrial contact.

Celestial Chariots and Cosmic Warfare: The Mahabharata's Vision of the Supernatural

The *Mahabharata*, a monumental epic composed between 400 BCE and 400 CE, tells the story of a devastating conflict between two branches of a royal family. Yet within its narrative lies more than just family drama and epic battles. The text presents detailed descriptions of supernatural beings, advanced weapons, and flying vehicles known as *vimanas*, which serve as divine chariots for gods and heroes alike. These descriptions are so vivid and intricate that they continue to inspire questions about their origins.

Vimanas: Divine Chariots or Ancient Aircraft?

In the *Mahabharata*, *vimanas* are depicted as flying vehicles used by gods and heroes, designed with a level of technological sophistication unimaginable in the ancient world. Unlike traditional human chariots, which rely on horses or oxen, these *vimanas* are said to fly through the skies independently, capable of traveling across vast distances at incredible speeds.

- **The Design and Capabilities of Vimanas**: The text describes *vimanas* as capable of moving in any direction, hovering, and even ascending to the heavens. They are described with intricate detail, sometimes including references to compartments, portals, and beams of light. Such features bear a resemblance to modern aircraft, with some enthusiasts speculating that these details could be based on

sightings of advanced flying machines beyond human technology at the time.
- **Weapons of Mass Destruction**: The *Mahabharata* also introduces weapons associated with *vimanas* that seem eerily similar to descriptions of modern armaments. The *Brahmastra*, for example, is a divine weapon that emits an intense, blinding light and heat that incinerates its target—a description that has drawn comparisons to nuclear weaponry. In the epic, the *Brahmastra* and other divine weapons are depicted as having the power to annihilate cities and armies. This narrative detail has led to speculation about whether these descriptions reflect a distant memory of advanced technology or simply serve as literary symbols of divine retribution.

The Kurukshetra War: A Battlefield Beyond Earthly Bounds

Central to the *Mahabharata* is the Kurukshetra War, a vast, cataclysmic conflict that pits two branches of a royal family against each other. Yet this is no ordinary war: it involves gods, demons, and heroes who wield supernatural powers and divine weapons that transform the battlefield into a cosmic arena.

- **Divine Intervention and Supernatural Beings**: Throughout the war, deities such as Krishna intervene, using *vimanas* to traverse the battlefield and aid their chosen heroes. These interventions present the gods as tangible, powerful figures who directly influence human affairs. The gods in the *Mahabharata* do not merely observe—they actively participate, aligning themselves with mortal causes and wielding powers that transcend human capabilities.
- **Symbolism or Reality?** The sheer scale and supernatural elements of the Kurukshetra War lead many scholars to interpret it as a symbolic struggle between good and evil. However, some researchers

argue that the vividness of the narrative and the level of detail in descriptions of technology suggest that these accounts could be based on real encounters with unknown beings. The possibility of ancient people witnessing advanced technology, whether of human or otherworldly origin, provides a provocative lens through which to view these stories.

Divine Rescue and Aerial Voyages: The Ramayana's Otherworldly Landscape

The *Ramayana*, another ancient Indian epic dating back to around 500 BCE, tells the tale of Prince Rama's quest to rescue his wife, Sita, from the clutches of the demon king Ravana. While this epic is filled with themes of love, loyalty, and dharma (righteous duty), it also presents characters who wield supernatural powers and use technology beyond the reach of mortal men. The *Ramayana* introduces a fascinating piece of technology known as the *Pushpaka Vimana*, which plays a pivotal role in Rama's journey and reveals more about ancient conceptions of divine power.

The Pushpaka Vimana: Ravana's Flying Fortress

The *Pushpaka Vimana*, a celestial chariot that originally belonged to the god of wealth, Kubera, is seized by Ravana and becomes a symbol of his power. Described as magnificent and self-propelled, the *Pushpaka Vimana* is able to fly vast distances and accommodate multiple passengers, demonstrating abilities far beyond the technology of the time.

- **A Technological Marvel**: The descriptions of the *Pushpaka Vimana* emphasize its autonomy and luxury, depicting it as a self-navigating vehicle that can move across landscapes and through the heavens with ease. Its ability to carry Rama, Sita, and others across vast distances suggests a vehicle of extraordinary design, hinting at capabilities that resemble modern aircraft or

even spacecraft. The detailed account of its features, including seating arrangements and propulsion, raises the question of whether this was purely mythological or based on an observed phenomenon.
- **A Journey Beyond Human Reach**: After rescuing Sita, Rama uses the *Pushpaka Vimana* to return home, marking the culmination of his journey with a flight that transcends earthly boundaries. In this instance, the *Pushpaka Vimana* not only serves as a vehicle but also represents Rama's triumph over evil and his return to his rightful place. The chariot thus functions both as a technological marvel and a symbol of divine favor.

A Tapestry of the Celestial and the Mundane: Interpreting Divine Chariots

The *Mahabharata* and *Ramayana* are not isolated in their portrayal of advanced vehicles and god-like beings. These epics reflect a broader cultural context in which humans perceived the sky as a realm of power, populated by beings who could shape the course of events on Earth. In interpreting these texts, scholars often debate whether the descriptions of *vimanas* and divine weapons are metaphors for spiritual concepts or symbolic representations of an advanced technology.

Symbolism, Memory, or Reality?

The stories of divine chariots, cosmic battles, and powerful weapons in the *Mahabharata* and *Ramayana* can be viewed from several perspectives:

1. **Symbolic Interpretation**: Many scholars argue that these elements are metaphorical, representing abstract spiritual concepts such as the struggle between good and evil, the pursuit of righteousness, and the transience of human life. In this view, the supernatural

elements serve to heighten the moral and philosophical themes of the epics.
2. **Cultural Memory of Advanced Technology**: Some proponents of ancient astronaut theories suggest that these descriptions are memories of a time when humanity encountered beings with advanced technology, possibly extraterrestrial in origin. They interpret the detailed descriptions of *vimanas* and powerful weapons as evidence that ancient people witnessed phenomena they could not fully comprehend.
3. **Mystical Experience**: Another interpretation suggests that the authors of these texts might have had mystical experiences, through meditation or altered states of consciousness, which they articulated in mythological language. In this framework, the divine chariots and cosmic battles are expressions of inner, spiritual revelations rather than literal encounters.

Bridging Ancient Myth and Modern Curiosity

The *Mahabharata* and *Ramayana* offer more than heroic stories; they present a vision of a world where human and divine realms intersect, where celestial beings interact with humanity, and where advanced technology is at the service of gods and heroes. Whether one views these elements as mythological symbolism, cultural memory, or allegory, the richness of these narratives continues to captivate modern audiences and raise questions about the nature of ancient encounters.

In exploring these epics, we gain insight into how ancient civilizations understood their relationship with the cosmos and the unknown. These stories, though thousands of years old, resonate with our modern quest for answers to fundamental questions about our origins, the possibility of extraterrestrial life, and the boundaries of human experience. As we continue this journey, the ancient Indian epics invite us to consider the profound impact of these narratives on human thought, bridging the gap between myth and reality in the ever-expanding story of humanity's place in the universe.

Prophetic Visions and Encounters in Religious Contexts: The Enigmatic Vision of Ezekiel

The Book of Ezekiel, one of the Hebrew Bible's most mysterious texts, contains a vision that has intrigued, puzzled, and inspired readers for centuries. Written by the prophet Ezekiel during his exile in Babylon around 593–571 BCE, this vivid and complex description of "wheels within wheels" has been interpreted in various ways: as a mystical experience, a revelation from God, and, more recently, a potential encounter with advanced technology. This vision—a detailed description of a divine chariot and its unusual occupants—provides a rare window into the ancient mindset and its relationship with the heavens. As we explore Ezekiel's vision, we encounter an intricate narrative that blends theology, prophecy, and perhaps even glimpses of an ancient understanding of the supernatural.

The Setting of the Vision: Babylonian Exile and a Prophetic Calling

Ezekiel's vision occurs against a backdrop of profound turmoil and upheaval. In 593 BCE, Ezekiel, a Jewish priest and prophet, found himself in exile along with other members of the Judahite community, deported to Babylon after the fall of Jerusalem. The Jewish people, grappling with the destruction of their homeland and the loss of their temple, faced a crisis of faith. It is within this context of despair and longing for divine guidance that Ezekiel receives his vision—a moment that would become foundational to his prophetic mission and provide spiritual reassurance to the exiled community.

The Scene of the Vision: By the River Chebar

The account opens with Ezekiel beside the river Chebar in Babylon. It is here, in a foreign land, that the heavens open, and Ezekiel encounters a supernatural spectacle. The river setting is significant: rivers in the ancient Near East were often associated with divine revelation and communication. The symbolism of water as a boundary between realms—the earthly and the divine—sets the stage for an encounter that transcends ordinary experience.

"As I was among the exiles by the river Chebar, the heavens were opened, and I saw visions of God." *(Ezekiel 1:1)*

The opening lines suggest that Ezekiel's vision is not simply an observation but a profound spiritual experience. The phrase "the heavens were opened" implies that Ezekiel is granted access to a hidden realm, where divine mysteries unfold before him.

Wheels, Fire, and Living Creatures: The Elements of Ezekiel's Vision

Ezekiel's vision unfolds with an elaborate and detailed description of the "glory of the Lord" manifested in the form of a chariot. This chariot is no ordinary vehicle; it is a structure of immense complexity, made up of interconnected wheels, fire, and supernatural beings. The imagery used in Ezekiel's account has often been compared to descriptions of advanced machinery or even spacecraft, fueling speculation about the vision's possible extraterrestrial implications.

The Living Creatures: Beings of Power and Mystery

Ezekiel first describes four "living creatures," each with four faces and four wings, who accompany the chariot. The creatures' faces resemble those of a man, a lion, an ox, and an eagle, symbolizing various attributes: intelligence, strength, servitude, and majesty. These beings possess the uncanny ability to move in any direction without turning, an ability Ezekiel emphasizes as supernatural.

"Each of the creatures went straight forward; wherever the spirit would go, they went, without turning as they went." (Ezekiel 1:12)

The creatures are not merely symbolic representations; they are animated with life and purpose, acting as guardians or attendants of the divine chariot. Their faces reflect characteristics that ancient cultures revered, yet their description defies conventional imagery. This combination of human and animal traits mirrors depictions found in Babylonian and Mesopotamian art, suggesting Ezekiel may have drawn on familiar symbols while imbuing them with a uniquely Hebrew meaning.

The Wheels Within Wheels: An Unprecedented Mechanism

Perhaps the most intriguing aspect of Ezekiel's vision is his description of the "wheels within wheels." These wheels are said to move alongside the living creatures, turning in all directions and emanating light. Each wheel contains an inner wheel, allowing movement on multiple axes, a feature unparalleled in ancient literature.

"Their appearance and construction was something like a wheel within a wheel… when they moved, they went in any of the four directions without veering as they moved." (Ezekiel 1:16-17)

The complexity of this structure has led to diverse interpretations. The wheels' design—moving independently and glowing with a strange radiance—suggests a sophisticated mechanism. For ancient readers, this vision would have evoked awe, associating such complexity with divine craftsmanship. However, modern interpretations have compared the description to rotating machinery or even spacecraft, with some theorists suggesting that Ezekiel might have witnessed technology beyond his comprehension.

The Chariot and the Throne: A Heavenly Vehicle

The living creatures and wheels form the base of the divine chariot, atop which rests a platform supporting a radiant throne. Upon this throne sits "the likeness of a man," surrounded by a rainbow-like glow, symbolizing divinity and authority. This figure, understood to be a representation of God, speaks to Ezekiel, commissioning him as a prophet to Israel.

The chariot's design and movement symbolize divine omnipresence and mobility. Unlike the idols of Babylon, bound to their temples, this image of God is mobile and transcendent, reflecting a deity who is not limited by earthly constraints. The chariot, therefore, serves as a theological statement: it portrays a God who accompanies the people in exile, capable of reaching them even in foreign lands.

Interpretations of Ezekiel's Vision: Divine, Symbolic, or Extraterrestrial?

Ezekiel's vision has been the subject of intense debate and varied interpretations, each seeking to understand the nature of this enigmatic experience. These interpretations can be broadly categorized into theological, symbolic, and extraterrestrial perspectives, each offering unique insights into the vision's meaning.

Theological and Symbolic Interpretations: A Vision of God's Glory

For traditional scholars and theologians, Ezekiel's vision represents a profound revelation of God's presence and glory. The chariot embodies divine sovereignty, while the creatures and wheels signify the attributes of God's creation. In this view, the vision is not meant to be understood literally but as a symbolic representation of God's power and majesty.

In Jewish mysticism, particularly within the tradition of **Merkabah** (chariot) mysticism, Ezekiel's vision is interpreted as a journey into divine mysteries. This tradition, which flourished in later centuries, viewed the chariot as a path to spiritual ascension, with each element of the vision representing stages of enlightenment. The complexity of the vision, in this interpretation, reflects the transcendent nature of divine reality, which cannot be easily comprehended by human minds.

Modern Speculations: Advanced Technology and Extraterrestrial Hypotheses

In recent years, Ezekiel's vision has attracted the attention of ancient astronaut theorists, who propose that the vision might describe an encounter with advanced extraterrestrial technology. Proponents of this theory suggest that the "wheels within wheels" and the luminous chariot could represent spacecraft, with the creatures serving as pilots or operators.

- Erich von Däniken and other proponents of the ancient astronaut theory have argued that Ezekiel's vision resembles descriptions of space travel, with the wheels possibly functioning as rotating parts of a spacecraft. They point to the vision's detailed mechanical elements as evidence that Ezekiel may have encountered an advanced civilization capable of interstellar travel.
- **Technical Analyses**: Some researchers have conducted technical analyses, translating Ezekiel's descriptions into engineering terms. They suggest that the wheels' multidirectional movement could reflect a form of anti-gravity technology, with the creatures controlling the chariot's direction and speed. While speculative, these analyses attempt to provide a

modern framework for understanding the vision's mechanical elements.

Ezekiel's Encounter in Cultural Context: A Reflection of Ancient Near Eastern Symbolism

While Ezekiel's vision is unique in its complexity, it also shares symbolic elements with Near Eastern art and literature. The creatures with animal-human features, the radiant throne, and the concept of a divine chariot echo themes found in Babylonian and Mesopotamian iconography. For example, the Lamashtu and Apkallu figures in Mesopotamian art also feature hybrid creatures, symbolizing the fusion of divine and earthly attributes.

Ezekiel's imagery, though distinct in its theological message, likely drew from the cultural symbols familiar to his audience. The similarities between the vision and Near Eastern motifs suggest that Ezekiel employed known symbols to communicate a radically new vision of God's sovereignty—a God who transcends human limitations and accompanies His people in exile.

Ezekiel's Vision as a Window into the Divine and the Unknown

Ezekiel's vision of wheels within wheels, radiant creatures, and a divine chariot serves as one of the most complex and enigmatic encounters in ancient religious literature. Whether viewed as a mystical experience, a symbolic vision, or a potential encounter with advanced technology, Ezekiel's account transcends simple interpretation. It reflects a profound moment of revelation, one that reassures a displaced people of God's presence while challenging the boundaries of human understanding.

In exploring Ezekiel's vision, we gain insight into the ancient mind's capacity to blend the spiritual with the physical, creating a narrative that is at once deeply symbolic and richly imaginative. For modern readers, the vision continues to provoke questions about the nature of divine encounters, the possibility of ancient technological knowledge, and the ways in which humans interpret the unknown. As we move forward, Ezekiel's vision reminds us of the enduring power of visionary experiences to inspire, mystify, and connect us to realms beyond our understanding.

The Mystery of "Flames in the Sky": Plutarch's Account of the Battle of Pharsalus

In 48 BCE, the fate of the Roman Republic hung in the balance as two powerful leaders—Julius Caesar and Pompey the Great—faced each other in the Battle of Pharsalus. This battle, one of the most consequential in Roman history, marked the turning point in Caesar's rise to power and the beginning of the end for Pompey. Amidst the turmoil of war, a strange phenomenon was recorded by the Greek historian Plutarch: an appearance of "flames in the sky." For centuries, this celestial event has been interpreted and reinterpreted, leaving us to wonder whether it was a divine sign, an atmospheric anomaly, or, as some suggest, an encounter with the unknown.

This account presents a fascinating blend of historical context, mythological interpretation, and supernatural speculation, offering insights not only into the events of that day but also into the ways ancient civilizations viewed the relationship between the heavens and earthly events. In delving into Plutarch's record of "flames in the sky," we explore the symbolism, cultural significance, and potential implications of this extraordinary event.

Setting the Scene: The Battle of Pharsalus and the Tumult of the Late Roman Republic

The Battle of Pharsalus was not simply a clash of armies; it was the culmination of a long-standing political and personal rivalry. The stakes were high, with Rome's future hanging in the balance. Caesar and Pompey, once allies, had become fierce adversaries, each supported by loyal factions vying for control of the Republic.

The Tension and Desperation of Civil War

The late Republic was a period of political instability, power struggles, and social unrest. The civil war between Caesar and Pompey represented more than a personal vendetta—it was a battle for Rome's future. Caesar, representing the populist cause, sought to consolidate power and reform the Republic, while Pompey and his supporters championed the traditional aristocratic order. When their armies met on the fields of Pharsalus, it was not only soldiers who witnessed the events of that day; the whole of Rome awaited the outcome, knowing it would shape their world.

Plutarch's Role as Historian and Storyteller

Plutarch, writing in the first century CE, was not just a historian but also a moralist and storyteller. His work, especially *Parallel Lives*, aimed to illuminate the virtues and vices of historical figures by comparing them across Greek and Roman history. His recounting of the "flames in the sky" during the Battle of Pharsalus thus serves multiple purposes: to record the events, inspire moral reflection, and offer a glimpse of the divine influence on human affairs.

The Phenomenon: "Flames in the Sky" During the Battle

As Caesar and Pompey's forces clashed, Plutarch describes a moment when strange lights appeared in the sky, resembling flames. He recounts that these "flames" were visible to many and interpreted as an omen by those who witnessed the battle. The appearance of these celestial lights, seemingly unrelated to the physical confrontation on the ground, added an otherworldly dimension to an already intense battle.

What Did Plutarch Actually Describe?

The "flames" Plutarch describes do not appear to have a natural explanation readily available in the historical record, such as lightning or meteor activity. Instead, his account suggests a glowing or fiery phenomenon with a sustained presence. For ancient observers, such an event would have defied natural explanation, reinforcing the belief that the gods intervened in human affairs.

Plutarch's account does not give precise details of the lights' duration, shape, or color, which has fueled speculation about their nature. Modern interpretations range from the mundane to the extraordinary, including:

- **Meteor Showers**: One plausible natural explanation is a meteor shower, which could create a series of bright, fast-moving lights. However, meteor showers are typically visible at night, and Plutarch does not indicate that the battle took place after dark.
- **Electrical Phenomena**: Atmospheric events like St. Elmo's fire, a phenomenon where plasma creates a glowing appearance around objects during electrical storms, could account for "flames." Yet this would typically occur in association with thunderstorms, which Plutarch does not mention.
- **Extraterrestrial Hypothesis**: Some theorists propose that Plutarch's "flames" could represent an early

record of a UFO sighting. The idea is that an extraterrestrial craft might have hovered or flown over the battlefield, observed by both armies as an unexplained fiery display.

Each theory carries its limitations, yet the lack of concrete detail leaves room for open interpretation, allowing each era to infuse the account with its understanding and worldview.

Ancient Interpretations: Omens from the Gods

For the ancient Romans, celestial phenomena were often interpreted as signs from the gods. Rome, heavily influenced by augury—the practice of interpreting omens from natural signs—would have seen such an event as bearing significant meaning. The appearance of flames in the sky over the battlefield would not have been dismissed as mere coincidence; rather, it would have been interpreted as a divine message, a sign of favor or disapproval from the gods themselves.

Divine Intervention in Times of Conflict

In Rome's religious worldview, gods and spirits were actively engaged in human events. Mars, the god of war, was believed to influence battles directly, and celestial phenomena were commonly seen as manifestations of divine will. For the soldiers and generals on the battlefield, the flames could have signaled victory or impending doom, depending on their interpretation.

The flames in the sky would have served as an emotional and psychological catalyst, potentially influencing the morale of Caesar's and Pompey's troops. The belief that the gods had taken an interest in the battle could have intensified the soldiers' convictions, with each side hoping to be favored by the divine forces observing them from above.

A Timeless Mystery: Historical Interpretations and Modern Theories

While Plutarch's account does not offer a clear explanation of the "flames in the sky," the story has persisted, sparking fascination and speculation through the centuries. Each era, informed by its own understanding of the universe, has brought new interpretations to the story. In the modern era, as UFO sightings and extraterrestrial theories have gained popularity, the story of the "flames" at Pharsalus has been reexamined from a different perspective.

A Possible UFO Encounter?

With the rise of UFO research, Plutarch's account has sometimes been interpreted as a possible early sighting of an extraterrestrial craft. The description of fiery lights moving independently across the sky resembles some accounts of UFO sightings, particularly those involving unexplained lights or crafts that appear to defy known physical laws.

While ancient societies lacked the technological understanding to recognize extraterrestrial visitors, the appearance of "flames" over a critical moment in history raises the possibility that ancient people might have encountered phenomena they could not explain. For proponents of the ancient astronaut theory, Plutarch's account offers evidence of unidentified aerial phenomena witnessed during a decisive battle—events that could have influenced the course of history.

Symbolism and Significance: Why Did Plutarch Include the Story?

The inclusion of the "flames in the sky" in Plutarch's account serves a symbolic function, elevating the Battle of Pharsalus from a mere historical event to a moment of cosmic importance. By emphasizing this extraordinary phenomenon,

Plutarch imbues the battle with a sense of divine presence, suggesting that the conflict between Caesar and Pompey was not only a human struggle but also one that resonated with the gods. The flames become an enduring symbol of fate's involvement in history, as if the universe itself bore witness to the turning point of Rome's destiny.

A Mystery That Endures Through Time

Plutarch's account of "flames in the sky" during the Battle of Pharsalus is a tantalizing fragment of history that offers a glimpse into how ancient civilizations interpreted the unknown. The appearance of fiery lights over a battlefield filled with tension and bloodshed would have been seen as a message from the divine, adding a layer of meaning to an already monumental event. Whether the flames represented an atmospheric phenomenon, an omen from the gods, or perhaps an extraterrestrial encounter, the story serves as a reminder of humanity's age-old desire to find meaning in the skies.

As we look back on this account, we see how each interpretation—ancient or modern—reveals more about our own cultural frameworks than about the event itself. The flames remain a mystery, but one that continues to ignite our curiosity, linking us across centuries to those who first looked up at the sky and wondered what forces might be watching.

Celestial Sightings over Rome: The Annales of Tacitus and a Mystery in the Sky (98 CE)

The historian Tacitus, revered for his meticulous documentation and profound insight into Roman life, wrote the *Annales*, a comprehensive historical record of the Roman Empire. Within this expansive work, Tacitus describes a series of unusual celestial phenomena witnessed over Rome—a sighting that has intrigued scholars and modern enthusiasts alike. The account, marked by Tacitus's characteristic detail, offers a rare glimpse into how ancient Romans perceived unexplained aerial phenomena and how they interpreted these occurrences through the lens of cultural beliefs and religious symbolism. As we explore Tacitus's account, we delve into the historical, cultural, and interpretative dimensions of these celestial events, examining their possible meanings, implications, and enduring fascination.

Tacitus and His *Annales*: The Voice of Rome's Past

Cornelius Tacitus, a senator and historian of the early Roman Empire, lived during one of the most turbulent periods in Roman history. His works, including the *Annales*, document events from the reign of Augustus to that of Nero. Tacitus was known for his dedication to historical accuracy and his critical approach to the moral character of Rome's leaders. His works often blend factual reporting with reflections on Rome's values, societal dynamics, and divine interventions in earthly affairs.

The *Annales* is a significant historical record, providing detailed descriptions of Rome's political intrigues, social norms, and cultural practices. Tacitus's attention to detail and his quest to understand not only the events but also their greater implications make his accounts invaluable for historians. Thus, when Tacitus records unusual celestial sightings, we are presented not just with a factual description but with a narrative steeped in cultural symbolism and philosophical inquiry.

The Celestial Event: "Unusual Sightings" over Rome in 98 CE

In 98 CE, during the reign of Emperor Nerva, Tacitus documented an extraordinary celestial phenomenon witnessed by the people of Rome. The nature of these "unusual sightings" was both striking and mysterious, leading to speculation and interpretation. According to Tacitus's description, the event was characterized by bright lights that appeared in the sky, resembling what we might now describe as fiery orbs or possibly even metallic objects reflecting sunlight.

Describing the Indescribable: Tacitus's Account of the Lights

Tacitus's description is evocative yet leaves room for interpretation. The lights in the sky were visible to many and were described as moving in ways that defied typical celestial patterns. This movement—seemingly intentional and coordinated—would have been especially unsettling to ancient observers, who understood stars and planets to follow predictable courses.

Tacitus's account is brief but vivid, highlighting several key aspects:

1. **Brightness and Color**: The lights were reported as unusually bright and potentially multicolored, giving them a visual impact beyond that of ordinary celestial bodies.
2. **Motion and Formation**: The movement of the lights, which seemed coordinated, was unprecedented. Ancient Rome, grounded in astrology and familiar with planetary patterns, would have perceived this as highly irregular and possibly threatening.
3. **Duration and Visibility**: The sighting was reportedly visible over a considerable period, witnessed by many, suggesting that it was not a fleeting meteor or other transient phenomenon.

Tacitus refrains from offering a concrete interpretation, leaving the mystery intact. This restraint may reflect his role as a historian rather than an augur or priest, yet his choice to include the event in the *Annales* implies that he deemed it significant.

Rome's Relationship with the Skies: The Cultural Context of Celestial Signs

In ancient Rome, celestial phenomena were often interpreted as messages from the gods, especially during periods of political or social instability. Augury—the practice of interpreting omens based on natural phenomena—was deeply ingrained in Roman society. A sighting as unusual as the one described by Tacitus would not have been taken lightly; it would have sparked discussions among augurs, priests, and the public, each eager to understand its implications.

Divine Omens: Signs of Favor or Displeasure

For Romans, the appearance of unusual celestial lights often signaled a message from the gods, either as a warning of misfortune or as a symbol of divine intervention. During periods of political upheaval, such signs were closely scrutinized. The reign of Nerva, though generally stable, marked a transition in Roman leadership, following the violent and unpredictable rule of Emperor Domitian. A celestial phenomenon like the one Tacitus describes might have been interpreted as an indication of either divine support or a potential warning of future turbulence.

- **Interpreting Favorable vs. Unfavorable Omens**: Romans believed that bright, steady lights in the sky might signify favor from the gods, while chaotic or disruptive patterns could indicate divine displeasure. Given the coordinated movement of the lights, ancient observers might have speculated that the gods were involved in a cosmic alignment, perhaps as a gesture of approval for Nerva's rule or as a harbinger of things to come.

Astrology and the Influence of Stars

Astrology held a respected place in Roman society, particularly among the elite. The alignment and movement of stars and planets were believed to impact events on Earth, from the success of military campaigns to the fates of emperors. Astrologers would have likely been consulted to interpret the unusual movement of the lights seen over Rome, using their knowledge of planetary behavior to predict future events.

Astrological interpretations, however, would have struggled to make sense of such an unexplainable phenomenon. Since the lights did not align with known planets or celestial bodies, astrologers may have resorted to symbolic interpretations, viewing the lights as markers of an impending shift in the cosmic order.

Theories and Interpretations: Ancient Explanations and Modern Speculations

Tacitus's description has led to numerous theories, both ancient and modern, about the nature of the lights. While ancient interpretations focused on divine or astrological explanations, modern readers have suggested a range of natural and supernatural causes, each reflecting the knowledge and beliefs of its time.

Natural Explanations: Atmospheric and Astronomical Phenomena

Some scholars argue that the lights over Rome might have been caused by natural atmospheric phenomena such as:

- **Auroras**: Although rare in Rome, auroras can occasionally appear at lower latitudes, especially during periods of intense solar activity. These colorful

lights could explain the bright and potentially multicolored appearance described by Tacitus.
- **Meteors and Comets**: While Tacitus does not explicitly describe the lights as shooting stars or comets, meteors could create a brief, intense display. However, their transient nature makes this explanation less likely given the duration of the sighting.
- **Electrical Phenomena**: St. Elmo's fire or other electrical events may have produced an unusual glow, although these phenomena are typically associated with thunderstorms, which Tacitus does not mention.

Each of these explanations has limitations, as none fully align with Tacitus's description of sustained, coordinated movement.

Extraterrestrial Theories: An Ancient UFO Encounter?

In modern times, Tacitus's account has been interpreted by some as a possible UFO sighting. Proponents of this theory point to the coordinated movement and brightness of the lights, suggesting that ancient people might have witnessed an unidentified flying object.

While the concept of extraterrestrial encounters was alien to ancient Rome, the behavior of the lights—as described by Tacitus—does resemble contemporary reports of UFO sightings, where witnesses often report patterns of movement that defy known physics.

Symbolic and Religious Interpretations

For many historians, the most plausible interpretation lies in Rome's cultural context. Tacitus's decision to include the sighting in his *Annales* may have been influenced by his awareness of the Roman public's interest in omens and signs. The unusual lights could thus be seen as a narrative device, enhancing the drama of his historical account and highlighting the precariousness of Rome's political climate.

Tacitus's Legacy: The Enduring Mystery of Rome's Celestial Lights

Tacitus's *Annales* is more than a historical record; it is a chronicle that captures the complexities of Rome's values, beliefs, and fears. His account of the celestial lights over Rome reflects an ancient fascination with the skies, an enduring curiosity that has spanned millennia. Whether a divine message, a natural phenomenon, or an encounter with the unknown, this sighting has retained its mystery, inviting each generation to reinterpret its significance.

A Celestial Puzzle Across Time

Tacitus's account of unusual celestial lights over Rome remains an enigma. It challenges us to consider how ancient civilizations interpreted the mysteries of the sky, blending theology, symbolism, and science in their search for meaning. For the Romans, the lights likely symbolized something beyond mortal comprehension, a reminder of the gods' omnipresence in both mundane and extraordinary affairs. Today, this ancient record of a celestial sighting serves as a window into Rome's cultural psyche—a society that saw the heavens not as a distant realm, but as an active, powerful force intertwined with human destiny. Tacitus's words have immortalized a moment of wonder, inviting us to continue exploring the boundaries between history, mythology, and the cosmic unknown.

The Cloudships of Magonia: Agobard of Lyon's Vision of the Unseen (9th Century CE)

In the 9th century, a mysterious and unusual belief took root in the minds of some communities in France: the idea of "cloudships" descending from a realm called Magonia. This belief posited that supernatural or otherworldly beings arrived in ships floating in the sky, influencing weather patterns and even abducting people. Agobard of Lyon, an influential bishop and theologian of his time, sought to address this strange phenomenon in his writings, both to denounce what he viewed as superstitious beliefs and to establish a framework for distinguishing between natural events and spiritual deception. His text, *De Grandine et Tonitruis* (*On Hail and Thunder*), provides an insightful and rare glimpse into medieval thought, revealing how ancient people interpreted unusual or catastrophic events as manifestations of supernatural intervention.

In this exploration of Agobard's account of cloudships and Magonia, we delve into the socio-religious context of medieval France, Agobard's interpretation and critique of these beliefs, and the modern significance of these narratives within discussions of extraterrestrial phenomena. We begin by immersing ourselves in the backdrop of Agobard's world—a time when the boundary between heaven and earth was seen as thin and permeable, and divine or supernatural forces were believed to influence daily life.

Agobard of Lyon: Bishop, Scholar, and the Voice of Rationality

Agobard, who served as the bishop of Lyon from 816 to 840 CE, was an important figure in the Carolingian Renaissance—a period of intellectual and cultural revival under the rule of Charlemagne and his successors. Known for his eloquent Latin prose and theological insight, Agobard addressed various issues of his day, from doctrinal debates to social justice concerns. He was particularly outspoken against superstitions and beliefs that, in his view, contradicted Christian doctrine and rational thought.

A World in Need of Explanation: Agobard's Mission

The 9th century was a time of both wonder and fear, as people faced unexplained natural phenomena with a mixture of awe and superstition. Events like hailstorms, thunder, and drought were often interpreted as divine punishment or the work of malevolent forces. In *De Grandine et Tonitruis*, Agobard expressed his concern about how easily people were led to believe in supernatural explanations for natural events. He saw it as part of his mission to redirect this fear towards a faith-based understanding of nature, grounded in God's providence rather than mythical narratives or pagan beliefs.

The Myth of Magonia: A Land Beyond the Clouds

One of the most unusual and striking aspects of Agobard's text is his description of the belief in "cloudships" from Magonia. According to this folklore, Magonia was a distant, mysterious realm in the sky inhabited by beings who could control weather patterns. It was believed that these beings would descend in their cloudships to collect crops destroyed by storms, which were supposedly gifts or offerings from people in exchange for manipulating the weather.

The Cloudships: Celestial Vessels or Apparitions?

The concept of cloudships is particularly intriguing because it portrays a vision of beings traveling through the sky in what seems to resemble ships or vessels—a portrayal not unlike modern interpretations of UFOs. In medieval art and literature, the sky was often depicted as the domain of angels, demons, and other supernatural beings. However, the idea of ships in the sky represented a more specific vision, suggesting that these beings possessed advanced, recognizable forms of transportation.

- **A Sense of Tangibility**: Unlike other supernatural beings who could manifest invisibly, the cloudships of

Magonia had a physical presence in people's minds, emphasizing a belief that these were real entities with a distinct form and purpose.
- **Connection to Weather Magic**: The people who believed in Magonia saw the cloudships as central to their understanding of weather phenomena. When storms or hail damaged crops, it was often assumed that the people of Magonia were involved, highlighting a perception of the natural world as manipulable by otherworldly forces.

The Role of Magonia's Beings in Medieval Cosmology

The inhabitants of Magonia were neither gods nor saints but were viewed as intermediaries with control over natural forces. Some scholars interpret these beings as figures akin to spirits or nature deities in pre-Christian belief systems, which saw nature itself as animated by conscious, willful beings. Although Christianity had by then become the dominant religion in Europe, older traditions and beliefs in nature spirits remained influential, and the people of Magonia reflect this synthesis of folklore and emerging Christian cosmology.

Agobard's Perspective: A Fierce Critic of Superstition

Agobard's writings on cloudships and Magonia reveal his strong stance against what he deemed superstitious or pagan beliefs. In *De Grandine et Tonitruis*, he condemns the credulity of those who believe in Magonia and the idea that storms could be influenced by otherworldly beings. For Agobard, the belief in cloudships not only contradicted Christian doctrine but also represented a form of spiritual deception that drew people away from true faith.

Faith vs. Folklore: Agobard's Arguments

In his critique, Agobard juxtaposes faith in divine providence with what he sees as the misguided belief in supernatural forces that act independently of God. He argues that attributing natural events to the actions of mythical beings diminishes the power and will of God, replacing divine order with superstition.

"People are so far deluded as to believe in the existence of a region called Magonia, whence come ships in the clouds."

By documenting and denouncing the belief in Magonia, Agobard sought to educate his followers, reinforcing the notion that all events in the natural world were under God's command and should be understood through the lens of Christian theology, not through pagan or fantastical narratives.

Agobard's Rationalism: Bridging Faith and Observation

In his rebuttal of the cloudship belief, Agobard demonstrates an early form of rationalism. His arguments emphasize the importance of distinguishing between reality and imagination, urging people to ground their beliefs in observable truths rather than myth. Agobard's skepticism toward popular superstition foreshadows the intellectual movement that would eventually shape the Renaissance and Enlightenment, as scholars began to question, investigate, and empirically analyze the natural world.

Modern Interpretations: Magonia and Extraterrestrial Speculation

While Agobard intended to debunk the belief in Magonia, his account has taken on new life in modern interpretations. The image of cloudships from another realm has captivated the imagination of UFO enthusiasts, paranormal researchers, and folklorists, sparking theories about ancient encounters with extraterrestrial beings.

The Ancient Astronaut Theory and Magonia

The idea of "cloudships" aligns with elements of the ancient astronaut theory, which suggests that extraterrestrial beings visited Earth in the past, possibly influencing human culture and mythology. Proponents of this theory interpret the story of Magonia as an early account of UFOs or alien intervention, positing that the beings from Magonia were not mythical figures but advanced entities who traveled in vessels through the skies.

- **Technological Interpretation**: The description of ships appearing from a distant realm suggests to some a vision of technology far beyond that of the 9th century. This interpretation leads to the suggestion that ancient people could have misinterpreted encounters with advanced beings or technologies, explaining them in terms familiar to their cultural context.
- **Nature Manipulation as Extraterrestrial Influence**: Some ancient astronaut theorists argue that the beings of Magonia may represent an early memory of extraterrestrials who possessed the means to alter or affect weather conditions, a concept popularized in modern UFO lore.

Folkloric and Psychological Interpretations: Magonia as Archetype

Others approach the story of Magonia from a folkloric or psychological perspective. The concept of cloudships reflects a common archetype—the mysterious, often threatening otherworldly visitors who have the power to influence or disrupt human life.

- **Cultural Reflection of Fear and Dependence on Nature**: In an agrarian society, dependent on the land for survival, the sudden destruction of crops by hail or storm could be devastating. The people of Magonia,

then, could be seen as an embodiment of these anxieties, a way to externalize and personify the unpredictability of nature.
- **Jungian Archetypes and the Collective Unconscious**: Psychologists like Carl Jung might interpret Magonia as a product of the collective unconscious, with the cloudships representing a manifestation of humanity's fear of the unknown and the otherworldly. Seen this way, the story of Magonia is less about specific beliefs and more about universal human experiences with forces beyond control.

Agobard's Legacy: The Persistence of the Unknown in Human Imagination

Although Agobard set out to disprove the existence of Magonia, his account has paradoxically helped to preserve it. The concept of cloudships from Magonia endures in folklore, literature, and even modern UFO theories, demonstrating the resilience of human imagination in the face of mystery. Agobard's writings offer a snapshot of a society struggling to balance faith and superstition, reason and imagination.

Magonia and the Question of What Lies Beyond

Agobard of Lyon's account of cloudships and the myth of Magonia illuminates the complexities of medieval thought, where superstition and rationalism coexisted, and unseen realms were believed to influence everyday life. While Agobard sought to ground his followers in faith-based explanations of natural events, his documentation of Magonia and cloudships has left us with a compelling mystery. Whether seen as a cautionary tale about the dangers of superstition, a precursor to modern extraterrestrial theories, or a psychological reflection of humanity's struggle with the unknown, the cloudships of Magonia continue to inspire curiosity and speculation, inviting each generation to reinterpret the mysteries of the sky and the stories they evoke.

The Nuremberg Celestial Event of 1561: A Fiery Battle in the Sky and Its Divine Significance

On April 14, 1561, the residents of Nuremberg, Germany, awoke to an extraordinary sight that would be etched in their memories—and in historical records—forever. As the morning sky over the city began to lighten, a series of unusual and alarming shapes appeared, prompting awe and fear among the townsfolk. What looked like an intense celestial battle unfolded before them, with spheres, crosses, and cylindrical objects seemingly locked in combat. This spectacle, now known as the 1561 Nuremberg Celestial Event, has since been a topic of fascination for historians, ufologists, and the general public, inspiring theories that range from divine warning to extraterrestrial encounter.

Exploring the Nuremberg Celestial Event requires understanding the socio-religious context of 16th-century Europe, examining the event's eyewitness accounts and woodcut illustration, and analyzing the interpretations, both historical and modern, that this event has inspired. By understanding the significance of the Nuremberg event to its contemporaries, we can better appreciate why it has retained such a powerful place in discussions of divine and extraterrestrial encounters.

Nuremberg in the 16th Century: A City Shaped by Religion, War, and Superstition

In 1561, Nuremberg was a prominent city within the Holy Roman Empire, known for its artisans, scholars, and merchants. However, it was also a time of intense religious and political upheaval. The Protestant Reformation, which had begun in 1517, had fractured the unity of Christianity across Europe, leading to violent conflicts and deep-seated mistrust between Catholic and Protestant states. Amidst these tensions, fear of divine retribution was prevalent, and natural or unusual phenomena were often interpreted as warnings from God.

A Worldview Steeped in Divine Omens

The worldview of 16th-century Europe was one in which the divine and supernatural played an active role in daily life. Signs in the sky—comets, eclipses, and meteor showers—were thought to be messages from God, foretelling significant events like wars, famines, or shifts in power. People in this era saw themselves as part of a cosmic struggle between good and evil, with celestial events often interpreted as symbols of this divine conflict. The Nuremberg event, seen as a vivid and violent display, would have struck a deep chord with the religious populace, who interpreted it as an omen with potentially apocalyptic implications.

The Event Unfolds: A "Battle" in the Sky

The morning of April 14, 1561, began like any other, until residents witnessed what seemed to be an extraordinary battle unfolding in the heavens. Eyewitnesses reported seeing red, blue, and black spheres, cylinders, crosses, and other shapes moving erratically across the sky, colliding with each other and producing a dazzling display.

The Eyewitness Account: A Moment of Fear and Wonder

A broadsheet published shortly after the event, created by local artist Hans Glaser, provides a detailed description of what the people saw. According to Glaser's broadsheet, the event took place around dawn and lasted for approximately an hour. The eyewitnesses described the following:

- **Spheres and Crosses**: Red and black spheres, along with crosses, appeared to "fight" each other in the sky. These objects moved quickly, interacting in a way that suggested deliberate motion and even a form of battle.

- **Cylindrical Objects**: Alongside the spheres, several cylindrical shapes appeared, some of which released additional smaller spheres. This detail adds to the impression that the objects had purposeful design or function.
- **Spectacular Colors and Light**: The shapes glowed with colors that were unusual and vivid. The red, blue, and black hues of the objects contrasted starkly against the morning sky, intensifying the otherworldly feel of the event.

The scene culminated in what appeared to be a fiery explosion, with objects seemingly "falling" from the sky and disappearing from view. This dramatic conclusion left the witnesses both captivated and terrified, feeling as though they had been privy to a divine or supernatural spectacle.

Hans Glaser's Woodcut: Visualizing the Celestial Encounter

Hans Glaser's woodcut illustration of the event, created soon after, is a remarkable piece of historical documentation. The woodcut not only visually captures the details of the eyewitness accounts but also provides insights into how the people of Nuremberg may have interpreted the event.

A Powerful Visual Record of the Phenomenon

In the woodcut, Glaser depicts a sky filled with crosses, spheres, and cylinders, each rendered with striking clarity. Some spheres appear to be "exploding" or breaking apart, while others clash with one another. The illustration reflects the sense of conflict and movement described by eyewitnesses, presenting a vivid impression of a chaotic, intense battle taking place above the city.

- **The Role of Artistic Interpretation**: Glaser's woodcut is not merely a literal depiction but also an interpretative piece. The arrangement of objects and

the intensity of the colors suggest that Glaser aimed to convey both the visual spectacle and the emotional impact of the event. As an artist of his time, Glaser's interpretation would have been shaped by his cultural and religious understanding, making the woodcut a blend of eyewitness account and artistic expression.
- **Symbolism of the Crosses and Spheres**: The use of crosses in the illustration would have had strong religious connotations for Glaser's contemporaries. In a deeply Christian society, the sight of crosses in the sky could easily be interpreted as a divine symbol, suggesting that the event had theological significance.

Interpreting the Event: Religious, Scientific, and Extraterrestrial Perspectives

Over the centuries, the Nuremberg Celestial Event has been interpreted in various ways. While 16th-century witnesses likely saw it as a divine warning or an apocalyptic sign, modern perspectives have introduced alternative explanations, ranging from natural phenomena to extraterrestrial encounters.

Contemporary Interpretations: A Divine Message?

For the people of Nuremberg, the celestial spectacle was likely perceived as a divine warning. The religious fervor of the time, combined with the symbolism of crosses and fiery lights, reinforced the belief that this was a message from God. The intensity and violence of the scene suggested that this was no ordinary omen but rather a sign of cosmic struggle—a heavenly battle between forces of good and evil.

- **Warnings of Impending Judgment**: Many in 16th-century Europe believed that signs in the sky foretold coming judgment, whether through plague, war, or famine. The unusual nature of the event and its dramatic display could easily be interpreted as an

omen, urging the people to repent or prepare for a forthcoming trial.
- **Apocalyptic Expectations**: The Protestant Reformation, along with frequent wars and natural disasters, contributed to a sense of impending apocalypse. Many saw the celestial event as a harbinger of the end times, a belief reinforced by the theological focus on celestial battles found in the Book of Revelation.

Scientific Theories: Atmospheric and Astronomical Explanations

Modern science offers potential explanations that attempt to demystify the event while recognizing the limits of 16th-century scientific knowledge.

- **Solar Phenomenon**: Some historians have suggested that the event might have been a rare atmospheric phenomenon, such as a sun dog or parhelion. Sun dogs are bright spots that appear on either side of the sun, caused by the refraction of sunlight through ice crystals in the atmosphere. However, sun dogs do not typically exhibit the movement or color described in the Nuremberg event.
- **Meteor Shower or Comet Fragmentation**: Another hypothesis is that the event was caused by a meteor shower or the breakup of a comet. Meteors or comet fragments burning up in the Earth's atmosphere could create bright lights and possibly even a sound, though these events are typically short-lived. The sustained and structured appearance of the objects described does not easily align with known meteor activity.

Extraterrestrial Hypotheses: An Early UFO Sighting?

The Nuremberg event has also attracted interest from UFO researchers, who see it as one of the earliest documented UFO sightings. The structured shapes, the behavior of the objects, and the use of terms like "battle" align with descriptions in modern UFO literature. Proponents of the extraterrestrial hypothesis argue that the event may represent an encounter with advanced technology or alien beings.

- **Structured Movement and Intentionality**: The reported movement of the spheres and cylinders—especially the way smaller objects emerged from larger ones—suggests coordinated action, an attribute often associated with intelligent control.
- **Symbolic Interpretation of Shapes**: Some UFO researchers speculate that the shapes might represent spacecraft, with the cylindrical objects possibly functioning as mother ships or launch platforms for smaller vessels. The appearance of cross shapes could reflect the witnesses' attempts to interpret unfamiliar forms in familiar, religious terms.

Legacy of the Nuremberg Event: A Story of Enduring Fascination

The Nuremberg Celestial Event remains an enduring mystery, its legacy preserved in Glaser's woodcut and the cultural memory of the event as a divine or supernatural phenomenon. The account captures a moment of profound wonder and fear, illustrating how communities throughout history have sought to understand and interpret the unknown.

A Battle in the Heavens, Real or Imagined?

The 1561 Nuremberg Celestial Event offers a window into the beliefs, fears, and imaginations of 16th-century Europeans. For the people of Nuremberg, the sight of strange objects in the sky was a powerful experience, interpreted through the lens of faith, cultural symbols, and the threat of divine judgment. While modern explanations propose natural phenomena or even extraterrestrial involvement, the event ultimately reminds us of humanity's enduring fascination with the skies and the mysteries they hold.

Whether as a divine warning, a rare atmospheric occurrence, or an encounter with the unknown, the Nuremberg Celestial Event continues to inspire curiosity and speculation. It invites us to consider the many ways we interpret extraordinary experiences and how these interpretations reflect our collective fears, hopes, and beliefs across the centuries.

The Fatima Apparitions of 1917: A Miraculous Encounter or an Extraterrestrial Phenomenon?

In the year 1917, during a turbulent time of war and political upheaval, three young shepherd children in the small village of Fatima, Portugal, claimed to witness a series of visions that would eventually captivate the world. This series of events, now known as the Fatima Apparitions, culminated on October 13, 1917, in what is famously referred to as the "Miracle of the Sun." In front of tens of thousands of people, the sun reportedly performed spectacular, inexplicable movements, leaving the crowd awestruck and fearful. While traditionally interpreted as a religious miracle, the event has sparked interest among UFO researchers who propose an alternative view: that the Miracle of the Sun may have been an encounter with a phenomenon more closely associated with extraterrestrial activity than divine intervention.

Exploring the Fatima Apparitions and the Miracle of the Sun requires an understanding of both the religious context and the characteristics of the phenomena witnessed. This investigation examines the historical setting, eyewitness testimonies, theological implications, and the modern interpretations that have brought extraterrestrial hypotheses into the conversation.

Setting the Stage: Portugal, 1917—A World at War and a Nation in Turmoil

The year 1917 was one of profound upheaval. Europe was embroiled in the devastation of World War I, and the traditional social structures of Europe were in turmoil. Portugal itself was undergoing significant political unrest, with tensions between the Catholic Church and a secular, anticlerical government that had gained power. In this environment of fear, uncertainty, and hope for divine intervention, religious faith was both a source of comfort and a focal point for interpreting extraordinary experiences.

Fatima: A Remote Village Drawn into Global Attention

Fatima, a rural town in central Portugal, was an unlikely location for a world-changing event. It was a simple village, largely isolated from the political and social turmoil of larger cities. However, the alleged encounters experienced by three young children—Lucia dos Santos and her cousins, Jacinta and Francisco Marto—would draw international attention to this remote place, turning it into a pilgrimage site for millions and a focal point for one of the 20th century's most famous religious phenomena.

The Apparitions: Visions of a Lady and Messages of Urgency

The story of the Fatima Apparitions begins in May 1917 when Lucia, Jacinta, and Francisco reported seeing a vision of a luminous lady who appeared above a small holm oak tree. This lady, who identified herself as "Our Lady of the Rosary," is widely regarded by the Catholic faithful as the Virgin Mary. Over the course of six months, the children reported seeing her on the 13th day of each month, receiving a series of messages that included warnings about repentance, the need for prayer, and ominous predictions about world events.

The Messages: Calls for Repentance and Visions of Catastrophe

The lady of Fatima allegedly shared with the children a series of messages that urged prayer, repentance, and devotion to God. These messages also contained prophecies about world events, including the end of World War I, the rise of communism, and the potential for a second, even more devastating conflict if humanity did not turn back to God.

- **Three Secrets of Fatima**: These messages included three "secrets," each with its own eschatological significance. The first described a vision of Hell, the second foretold the rise of communism and the persecution of Christians, and the third was a veiled prophecy about the struggles of the Church and potential global catastrophe.

The messages resonated with a Catholic populace already on edge from war and social unrest, reinforcing a sense of urgency and impending judgment. The culmination of these apparitions, however, was yet to come.

The Miracle of the Sun: A Spectacle Witnessed by Thousands

The final apparition, on October 13, 1917, drew an immense crowd—estimates range from 30,000 to 100,000 people—who had gathered to witness the promised miracle. Many were devout believers, while others were skeptics, reporters, and officials, drawn by the children's claims and the rising public interest in the events at Fatima. What followed was one of the most widely reported supernatural events of the 20th century, now known as the Miracle of the Sun.

The Event: The Sun "Dancing" in the Sky

Eyewitnesses of the Miracle of the Sun reported an astonishing series of events. After a period of rain, the clouds cleared, and the sun appeared as an unusual disc of dull silver. According to various accounts, the sun began to spin, emit radiant colors, and move erratically, plunging toward the earth in a terrifying manner before returning to its normal position. Witnesses described it as a "dance" of the sun, with colors shifting across the landscape and the crowd. This event left many in awe, with some fearing that the world was about to end.

- **Physical Effects**: Some witnesses reported physical effects such as warmth or dryness despite the rain-soaked ground, while others claimed miraculous healings. The phenomenon lasted approximately ten minutes, after which the sun returned to its usual state.

Eyewitness Accounts: A Diverse Range of Experiences

The accounts of the Miracle of the Sun vary, with some witnesses reporting clear and vivid descriptions of the sun's movement, while others reported seeing nothing unusual. Newspapers at the time recorded testimonies from those who claimed to have seen the phenomenon from miles away, while a few skeptics suggested mass hysteria or optical illusions as explanations.

Interpretation Through Faith: A Divine Miracle Confirming the Virgin Mary's Presence

For the Catholic Church and many of the faithful, the Miracle of the Sun was considered an authentic miracle—a sign from God confirming the message of the Virgin Mary. The Church officially recognized the apparitions at Fatima in

1930, lending credibility to the children's accounts and endorsing the events as divinely inspired. The Miracle of the Sun was thus enshrined as a manifestation of divine power, a call to repentance, and a reassurance of God's presence amidst a troubled world.

Theological Significance: Reinforcing Faith and Divine Protection

The Miracle of the Sun carried significant theological implications. It reinforced Catholic beliefs about the power of prayer, the intercession of the Virgin Mary, and the need for devotion and repentance. For believers, the phenomenon was not merely a spectacle but a spiritual confirmation, a tangible sign that heaven was attentive to humanity's struggles and willing to intervene in times of crisis.

Alternative Perspectives: Extraterrestrial and UFO Hypotheses

In modern times, the Miracle of the Sun has attracted the attention of UFO researchers who suggest that the event may not have been a divine intervention but rather an encounter with an extraterrestrial phenomenon. This interpretation posits that the "sun" seen by the crowd may have been a craft or object exhibiting advanced technology, creating the illusion of movement and emitting unusual light effects.

Possible Extraterrestrial Explanations for the Miracle of the Sun

Proponents of the extraterrestrial hypothesis point to several characteristics of the Miracle of the Sun that align with modern descriptions of UFO phenomena:

1. **Erratic Movement and Structured Light Patterns**: The sun's reported movement, as described by witnesses, does not correspond with known solar or

atmospheric behavior but aligns with accounts of structured, controlled movement often reported in UFO sightings.
2. **Color Changes and Visual Effects**: Witnesses reported seeing vibrant colors radiating from the "sun," a phenomenon that some researchers suggest could have been created by an energy field or propulsion system, similar to descriptions of electromagnetic interference associated with UFOs.
3. **Lack of Universal Observation**: Despite the large crowd, not everyone present saw the same phenomena, suggesting that the experience may have been perceptual or targeted, a phenomenon that some UFO researchers attribute to selective visibility or projection.

A Psychological and Optical Phenomenon?

Some skeptics and psychologists suggest that the Miracle of the Sun could have been the result of mass hysteria, a psychological phenomenon in which large groups of people influence each other's perceptions. Optical explanations, such as prolonged staring at the sun causing temporary retinal effects, have also been proposed, though these theories fail to account for the widespread consistency in witness descriptions and the duration of the event.

The Continuing Legacy of the Fatima Apparitions and the Miracle of the Sun

The Fatima Apparitions and the Miracle of the Sun continue to inspire devotion, theological reflection, and scientific curiosity. Fatima has become a significant site of pilgrimage, drawing millions of Catholics who see the events of 1917 as a profound manifestation of divine intervention. However, the event's characteristics have also intrigued secular and scientific researchers, keeping the discussion alive within the fields of ufology and paranormal studies.

Divine Intervention or Extraterrestrial Encounter?

The Fatima Apparitions and the Miracle of the Sun present a complex and multifaceted event that defies easy classification. For many, it remains a powerful affirmation of faith and a testament to divine involvement in human affairs. For others, it offers tantalizing clues about the possibility of advanced, otherworldly phenomena intersecting with human experience. Whether interpreted as a miracle, a mass psychological phenomenon, or an encounter with extraterrestrial technology, the events at Fatima stand as a testament to humanity's enduring fascination with the unknown, the divine, and the mysteries of the cosmos. The Miracle of the Sun remains a compelling reminder that our understanding of extraordinary phenomena is often shaped as much by belief and culture as by observation and science.

Religious Apparitions and Extraterrestrial Contact: Bridging Heaven and the Cosmos

Throughout history, humanity has sought answers to profound questions by looking to the skies, interpreting celestial events and supernatural experiences as signs from the divine. These religious visions—often of figures like angels, saints, and even the Virgin Mary—have long been regarded as proof of divine communication, offering messages of guidance, warning, and hope. In recent decades, however, these same apparitions have attracted interest from those who suggest that such encounters may not be purely spiritual in nature but could represent interactions with advanced extraterrestrial beings.

This cross-analysis of religious messages and modern interpretations of extraterrestrial contact explores the historical context, symbolic parallels, and evolving interpretations that link traditional religious experiences with contemporary UFO phenomena. By examining these encounters through multiple lenses, we gain a deeper understanding of how cultural and technological shifts shape our perceptions of the unknown and redefine our relationship with the cosmos.

A Timeless Human Need: Seeking Meaning in Extraordinary Experiences

From early civilizations to modern society, humans have consistently turned to the heavens for insight, purpose, and transcendence. Apparitions and celestial events have always been seen as means of communication from higher powers, especially during times of crisis or upheaval. In such encounters, people often report receiving messages about morality, repentance, impending danger, or global transformation.

The Historical Role of Apparitions in Religion

Religious apparitions have historically served as powerful tools for shaping belief systems and fostering communal identity. In the Christian tradition, apparitions of the Virgin Mary and various saints are often seen as direct interventions in human affairs, emphasizing messages of faith, devotion, and moral guidance. Similarly, other religions include accounts of spiritual encounters, from Hindu yogis experiencing divine visitations to Islamic prophets receiving messages from angels.

These experiences have several common themes:

- **Messages of Hope and Redemption**: Apparitions often bring messages that inspire people to seek redemption and moral reform, resonating with the spiritual and ethical needs of the time.
- **Warnings and Apocalyptic Themes**: Some apparitions warn of future events, especially during turbulent times, which may be interpreted as divine judgment or prophecy.

- **Visions of Higher Beings or Realms**: Beings in religious apparitions are often depicted as luminous, otherworldly figures who convey a sense of peace or awe, aligning with the idea of higher, transcendent realities.

Such themes are common across cultural boundaries, suggesting a universal pattern in how humans interpret unexplained phenomena as messages from beyond.

Comparing the Divine and the Extraterrestrial: Overlapping Characteristics

With the rise of UFO sightings and interest in extraterrestrial life in the 20th century, modern society has begun to notice intriguing similarities between historical religious apparitions and contemporary accounts of extraterrestrial contact. While religious apparitions are typically seen as interactions with divine or spiritual beings, UFO sightings are often interpreted as encounters with technologically advanced, possibly extraterrestrial life forms.

Similarities in Appearance and Setting

Both religious and extraterrestrial encounters frequently involve bright lights, hovering objects, and figures appearing in radiant, ethereal forms. For instance, witnesses of Marian apparitions often describe seeing a radiant figure surrounded by light, while those who report UFO sightings describe similar luminous objects or figures.

- **The Use of Light and Radiance**: Light is a common motif in both religious and UFO encounters. Apparitions of religious figures are often accompanied by blinding or colored light, which creates an awe-inspiring effect. UFO sightings, too, often include descriptions of glowing orbs, bright discs, or beams of

light, contributing to the perception of an encounter with something otherworldly.
- **Remote or Isolated Settings**: Both types of encounters are frequently reported in remote or quiet areas. Marian apparitions, such as those at Lourdes or Fatima, took place in secluded settings, as did many early UFO sightings. The secluded nature of these locations contributes to the sense of mystique and rarity, enhancing the impact of the experience.

Messaging and Intent: Parallels in Communication

The messages delivered during religious apparitions and alleged extraterrestrial contacts often convey similar themes, such as warnings about environmental degradation, calls for spiritual or moral transformation, and visions of apocalyptic scenarios. These messages have a universal quality, touching on aspects of human behavior, responsibility, and the fate of humanity.

- **Warnings About Humanity's Future**: Many Marian apparitions, including those at Fatima, involve warnings about future calamities, urging humanity to turn away from sin to avoid disaster. Similarly, messages associated with UFO encounters often include warnings about ecological destruction, nuclear proliferation, and other dangers facing humanity, suggesting that the extraterrestrial beings are concerned with the future of the planet.
- **Encouragement of Spiritual Growth and Unity**: Both religious and extraterrestrial messages sometimes emphasize the need for spiritual enlightenment, encouraging unity and compassion. This message is seen in apparitions like those at Medjugorje, which promote peace and prayer, as well as in reported telepathic communications from extraterrestrials, which advocate for harmonious coexistence.

From Angels to Aliens: Cultural Shifts in Perception

One of the most striking aspects of this cross-analysis is how cultural shifts and technological advancements influence interpretations of these experiences. In earlier centuries, when religious frameworks dominated cultural thought, people naturally interpreted extraordinary encounters through a theological lens. Today, in an age marked by science fiction, space exploration, and technological advancement, similar encounters are often reinterpreted as extraterrestrial rather than divine.

20th-Century Influence: UFOs and the Rise of the Extraterrestrial Narrative

The 20th century brought rapid technological progress and the popularization of space travel, reshaping humanity's imagination. As people became more familiar with the concept of advanced civilizations on other planets, the possibility of extraterrestrial life became mainstream. UFO sightings surged in the mid-20th century, particularly after events like the Roswell incident in 1947, leading to a shift in how extraordinary encounters were interpreted.

- **Reframing Apparitions as Extraterrestrial**: With increased exposure to the idea of alien life, some individuals began to reinterpret religious experiences as encounters with extraterrestrial beings. For example, the Miracle of the Sun at Fatima, long considered a miraculous event, has been re-examined by some ufologists as a possible UFO sighting, suggesting that the "sun" seen by the crowd could have been an extraterrestrial craft.
- **Symbolic Transition**: The transition from angels to aliens reflects a broader cultural shift. Angels, seen as intermediaries between humanity and the divine, are replaced in modern imagination by extraterrestrials—

beings who might possess advanced knowledge and technology. This shift reflects an evolution in humanity's understanding of higher beings, moving from a theological framework to a cosmological one.

Modern Interpretation: Reinterpreting Apparitions in Light of UFO Phenomena

Some modern UFO researchers suggest that apparitions and UFO sightings may be part of a single phenomenon, experienced differently depending on cultural context. According to this view, the beings encountered—whether called angels, saints, or extraterrestrials—might belong to the same class of entities, perceived differently based on the observer's beliefs and expectations.

- **John Mack's Hypothesis**: Dr. John Mack, a Harvard psychiatrist, theorized that encounters with UFOs and aliens may represent experiences that transcend our physical reality. He proposed that these encounters have spiritual dimensions, blurring the line between mystical and extraterrestrial experiences. Mack's approach suggests that extraterrestrial encounters may be spiritual or psychological events, rooted in a dimension beyond conventional perception.
- **Jacques Vallée's "Control System"**: Jacques Vallée, a prominent ufologist, has proposed that both religious and extraterrestrial encounters might serve as part of a "control system," influencing humanity's behavior and beliefs. According to Vallée, these encounters could be designed to guide human evolution, with messages adapted to fit the cultural context—appearing as divine revelations in religious eras and extraterrestrial visitations in modern times.

The Psychological Dimension: The Role of the Collective Unconscious

The Swiss psychologist Carl Jung explored the idea that UFO sightings might be manifestations of the collective unconscious. Jung suggested that these experiences tap into archetypal symbols, projecting inner psychological or spiritual conflicts onto external, otherworldly forms. In this sense, UFOs and religious apparitions alike may represent a projection of humanity's own fears, hopes, and longing for meaning.

- **The Visionary State**: Many religious apparitions involve witnesses experiencing altered states of consciousness, marked by intense visions and sensory phenomena. Similarly, UFO encounters often include reports of mental and emotional shifts, sometimes described as telepathic communication. These experiences could indicate a heightened visionary state, influenced by deep-seated archetypal images.
- **The Archetype of the "Higher Being"**: Both religious and extraterrestrial beings represent the archetype of the "Higher Being"—figures who possess knowledge, wisdom, and power beyond human understanding. The Virgin Mary, for example, embodies purity and divine wisdom, while extraterrestrial figures are often seen as bearers of advanced knowledge and concern for humanity's future. This archetype resonates across cultures, symbolizing humanity's search for guidance and connection with something greater.

The Unifying Quest: A Shared Longing for Connection Beyond Earth

At the heart of both religious and extraterrestrial encounters is a shared human desire to connect with a higher order, to understand the mysteries of existence, and to receive guidance. Whether people interpret these encounters as divine or extraterrestrial, they reveal a longing to transcend earthly limitations and to find meaning in the universe.

A Continuum of Belief: Bridging Religion and Cosmology

The cross-analysis of religious apparitions and UFO encounters suggests that both can be viewed as part of a continuum of belief, shaped by cultural context but rooted in universal human experience. This continuum bridges the sacred and the scientific, the spiritual and the extraterrestrial, allowing us to explore these encounters from multiple perspectives without dismissing their validity.

- **Beyond Labels**: Rather than strictly categorizing such experiences as either religious or extraterrestrial, this approach invites a more holistic understanding. Apparitions and UFO encounters may both reflect encounters with a transcendent reality, experienced in ways that resonate with individual beliefs and cultural frameworks.

A Mystery That Defies Easy Categorization

The cross-analysis of religious messages and modern interpretations of extraterrestrial contact reveals a complex and multifaceted phenomenon that transcends simplistic explanations. Throughout history, humanity has encountered mysterious beings and received messages that challenge our understanding of the universe. These experiences, whether described as divine apparitions or extraterrestrial contact, underscore a shared yearning to connect with something beyond ourselves.

In studying these encounters, we gain insight into not only the nature of these phenomena but also the ways in which human perception adapts to new ideas and frameworks. As our understanding of the cosmos evolves, so too may our interpretations of these experiences, allowing us to explore the unknown from ever-expanding perspectives. Ultimately, whether viewed through the lens of faith, science, or both, the messages embedded in these encounters invite us to question, reflect, and expand our sense of what it means to be human in a universe filled with mystery.

The Contactee Movement (1950s–1970s): Earth's Ambassadors to the Cosmos

In the 1950s, as the world grappled with Cold War anxieties, the fear of nuclear annihilation, and the dawn of the Space Age, a unique phenomenon emerged in the United States and soon spread worldwide. Known as the "Contactee Movement," this movement centered around individuals who claimed to have not only sighted UFOs but also established direct, personal contact with extraterrestrial beings. Unlike reports of alien abductions or terrifying encounters, the Contactee Movement was characterized by individuals who described their experiences as enlightening and benevolent. These "contactees" claimed that extraterrestrials had chosen them to convey messages of peace, unity, and spiritual advancement to humanity.

Central to this movement were key figures like George Adamski and Howard Menger, who became renowned for their claims of interaction with alien civilizations. The messages they received from these beings were often utopian and filled with calls for moral betterment, and their charismatic personalities attracted dedicated followers. To understand the Contactee Movement and its historical significance, we must explore the lives of these key figures, the messages they received, and the broader sociopolitical context that made these narratives resonate with so many people.

A New Phenomenon for a New Age: Post-War Anxieties and Cosmic Dreams

The Contactee Movement emerged in a period of profound technological and ideological transformation. The aftermath of World War II and the subsequent Cold War tension created a sense of global vulnerability and an urgent desire for peace. At the same time, advances in science and the nascent Space Age fueled the belief that humanity might not be alone in the universe.

The Cultural Backdrop: Fear of Nuclear Catastrophe and the Dawn of the Space Age

The bombings of Hiroshima and Nagasaki had left an indelible mark on the global consciousness, reminding humanity of its capacity for destruction. The fear of nuclear war became a recurring theme in popular culture and was accompanied by a growing awareness of the fragility of life on Earth. As people looked to the skies, the prospect of extraterrestrial life became a hopeful distraction, a sign that advanced civilizations might offer a different, peaceful way of life.

At the same time, the launch of Sputnik in 1957 and President John F. Kennedy's call to land a man on the moon by the end of the 1960s fostered a fascination with space exploration. It was within this context that the Contactee Movement emerged, appealing to both a longing for transcendence and an intense desire for world peace.

George Adamski: A Messenger from Venus

George Adamski, one of the most influential figures in the Contactee Movement, claimed that he had direct contact with extraterrestrials from Venus and other planets in our solar system. Born in Poland and raised in the United States, Adamski was an unusual figure, blending the roles of mystic, scientist, and self-proclaimed "cosmic philosopher." His experiences and teachings would come to define much of the Contactee Movement.

Adamski's First Encounter: A Meeting in the Desert

On November 20, 1952, Adamski claimed to have his first extraterrestrial encounter in the Mojave Desert near Desert Center, California. According to his account, Adamski and a group of friends spotted a large cigar-shaped object hovering in the sky. Shortly thereafter, Adamski claimed that a smaller scout ship landed, and he encountered a being named Orthon. Orthon was described as a tall, humanoid figure with long, flowing hair, dressed in a shiny one-piece suit. He communicated with Adamski telepathically, conveying messages about the dangers of nuclear weapons and the importance of peace on Earth.

This encounter would become one of the most famous stories in the Contactee Movement. Adamski wrote extensively about his meetings with Orthon and other extraterrestrial beings, describing their advanced technology, peaceful societies, and spiritual wisdom. His book, *Flying Saucers Have Landed* (co-authored with Desmond Leslie), became a bestseller and attracted significant public attention.

Messages from the "Space Brothers"

The beings Adamski encountered, whom he called "Space Brothers," were portrayed as enlightened and compassionate entities concerned with humanity's future. They warned of the self-destructive path that humanity was on, particularly in the context of nuclear warfare. The Space Brothers emphasized the importance of global unity, love, and the abandonment of warlike behavior.

Adamski's accounts had several recurring themes:

- **The Need for World Peace**: The Space Brothers repeatedly urged humanity to avoid nuclear weapons and to seek peaceful solutions to conflicts. They warned that Earth's violent tendencies could have consequences beyond our planet.
- **Spiritual Advancement**: Adamski's extraterrestrial contacts emphasized that humanity needed to evolve spiritually, advocating values such as compassion, humility, and harmony with nature. According to Adamski, the Space Brothers lived by these values, creating utopian societies that Earth could aspire to emulate.
- **Cosmic Brotherhood**: Adamski's messages often included a vision of a "cosmic brotherhood," a universal alliance of enlightened civilizations that operated in harmony. He described these beings as technologically advanced but also morally and spiritually superior, suggesting that humanity's entry into this cosmic alliance was contingent on adopting peaceful, ethical behavior.

Adamski's charismatic personality and detailed descriptions of his encounters drew a devoted following, with some believing he had been chosen as a messenger for humanity. However, his claims were also met with skepticism, as many scientists and members of the public questioned the feasibility of beings from Venus—a planet known to be uninhabitable by human standards.

Howard Menger: Contact in the Backyard

Another key figure in the Contactee Movement was Howard Menger, a sign painter from New Jersey who claimed a lifelong series of contacts with extraterrestrials. Unlike Adamski's desert encounters, Menger's experiences were closer to home. He described beings landing in his backyard and engaging in conversations with him, much like friendly neighbors visiting from afar.

Menger's Encounters: A Story of Familiarity and Friendship

Menger's encounters with extraterrestrials were different from Adamski's in tone and setting. He reported that beings who looked similar to humans visited him regularly, inviting him aboard their spacecraft and showing him glimpses of other planets. Menger claimed that these beings were highly advanced and benevolent, seeking to guide humanity toward a peaceful and prosperous future.

Menger's book, *From Outer Space to You*, published in 1959, documented his experiences and the messages he received. He described the extraterrestrials as peaceful, deeply compassionate beings who wanted to help humanity overcome its destructive tendencies. Like Adamski, Menger emphasized themes of universal love, environmental stewardship, and the importance of spiritual growth.

Messages of Unity, Healing, and Earth's Future

The messages Menger received were strikingly similar to those reported by Adamski. The extraterrestrials warned about environmental destruction, the dangers of nuclear weapons, and humanity's need for spiritual awakening. They also encouraged people to live in harmony with nature and each other, promoting values that would later resonate with the countercultural movements of the 1960s.

Menger described how the beings expressed sorrow over Earth's wars and environmental degradation, urging humans to abandon conflict and take responsibility for their planet. They introduced him to concepts of "universal harmony" and hinted that Earth could achieve a similar state if humanity embraced these values.

- **Healing Technologies**: Menger claimed that the extraterrestrials possessed healing technologies that could cure various diseases. He described the beings

as having perfected techniques that enabled them to live in radiant health, free from suffering, which they wished to share with Earth.
- **Environmental Awareness**: The beings Menger encountered were portrayed as caretakers of their own planets, living in perfect balance with their environments. This message struck a chord with those concerned about industrialization and environmental destruction, themes that would become central to the environmental movement of the 1970s.

The Legacy and Impact of the Contactee Movement

The messages delivered by Adamski, Menger, and other contactees reflected a blend of utopian ideals, spiritual beliefs, and humanitarian values. Their stories captivated audiences during a time of social uncertainty and have continued to resonate with those seeking alternative spiritual paths and a vision of a more peaceful world.

Influence on New Age and Countercultural Movements

The Contactee Movement played a significant role in shaping the New Age and countercultural movements of the 1960s and 1970s. The messages of universal love, spiritual advancement, and environmental responsibility appealed to those disillusioned with mainstream society, and the idea of extraterrestrial guidance aligned with the era's fascination with Eastern spirituality, mysticism, and alternative lifestyles.

Controversy and Criticism: Questions of Authenticity

While the Contactee Movement gained a substantial following, it was also subject to criticism and skepticism.

Many scientists and researchers argued that the claims made by Adamski, Menger, and others lacked evidence and were based on pseudoscience. Skeptics suggested that the messages received were reflections of human aspirations rather than genuine extraterrestrial communications.

Some contactees were accused of fabricating their stories for fame or financial gain, and the descriptions of advanced civilizations on Venus or Mars became increasingly implausible as scientists learned more about these planets. However, despite the controversies, the movement left a lasting impact on popular culture and the evolving narrative of extraterrestrial life.

A Movement Rooted in Hope and Cosmic Aspiration

The Contactee Movement of the 1950s and 1970s represents a unique chapter in the history of extraterrestrial encounters. Figures like George Adamski and Howard Menger, with their tales of benevolent Space Brothers, provided an alternative to the fear-based narratives that characterized much of the UFO discourse in the post-war period. Their messages of peace, unity, and spiritual growth offered a hopeful vision of humanity's potential future, one that resonated deeply with those seeking answers in a time of global uncertainty.

Whether viewed as a genuine phenomenon, a social movement, or a product of its cultural context, the Contactee Movement demonstrates humanity's ongoing fascination with the unknown and our deep-seated desire for guidance from beyond. In examining the lives and messages of these contactees, we gain insight into the collective psyche of the era and our universal longing for a world of peace, unity, and cosmic connection.

The Contactee Movement: Cold War, Nuclear Fears, and Messages from the Cosmos

The Contactee Movement of the 1950s and 1970s represents one of the most intriguing phenomena in the history of extraterrestrial narratives. This movement was fueled not only by claims of direct contact with extraterrestrials but also by a complex interplay of sociopolitical factors that shaped its development and appeal. Against the backdrop of the Cold War, as humanity confronted the specter of nuclear annihilation, contactees emerged with messages that conveyed profound warnings about human self-destruction, pleas for peace, and visions of a better world.

To understand the significance of the Contactee Movement, it is essential to examine the broader cultural context of the time, which was defined by global tensions, technological advances, and an undercurrent of existential dread. The interplay between the political climate, nuclear fears, and contactee narratives reveals the ways in which human anxieties and aspirations influenced interpretations of extraterrestrial encounters.

A World on the Brink: The Cold War and Heightened Existential Threats

The Contactee Movement arose in the shadow of the Cold War, a period of intense geopolitical tension between the United States and the Soviet Union. Following World War II, these two superpowers embarked on a political and ideological struggle that permeated nearly every aspect of society. This struggle was marked by the stockpiling of nuclear weapons, the development of long-range ballistic missiles, and the looming possibility of mutually assured destruction.

The Age of Nuclear Fear: A Constant, Unseen Threat

For much of the 1950s and 1960s, the fear of nuclear war was a pervasive element in everyday life. With the development and deployment of atomic bombs, the world had entered a new era in which a single conflict could result in catastrophic global consequences. The threat was ever-present, reinforced by publicized military drills, civil defense campaigns, and government-issued guidance on surviving a nuclear blast. For the first time, humanity had created weapons capable of destroying civilization as it was known, leading to widespread existential fear and moral questioning.

- **"Duck and Cover" Drills and Fallout Shelters**: In the United States, schools regularly conducted "duck and cover" drills, training children to seek shelter under their desks in the event of an attack. Families built fallout shelters in their basements or backyards, hoping to protect themselves from radiation. These measures reinforced the sense of an omnipresent threat that could strike without warning.
- **Cultural Reflections of Nuclear Fear**: The threat of nuclear war seeped into popular culture, with movies, books, and television shows exploring themes of atomic devastation and post-apocalyptic survival. From dystopian novels like *On the Beach* to science fiction films about radioactive monsters, nuclear anxieties influenced a broad spectrum of cultural expressions.

This environment of pervasive anxiety created fertile ground for narratives that offered either reassurance or warnings from a higher, wiser authority. The Contactee Movement emerged as a response to these anxieties, with contactees presenting extraterrestrial beings as messengers of peace who urged humanity to avoid self-destruction.

Contactees as Prophets of Peace: Messages Tailored to Cold War Anxieties

Against this backdrop of Cold War fears, contactees like George Adamski, Howard Menger, and others claimed to have received messages from extraterrestrials that directly addressed humanity's existential concerns. These messages often emphasized peace, warned against nuclear weapons, and urged humanity to embrace a more enlightened, compassionate way of life.

A Cosmic Warning: The Dangers of Nuclear Weaponry

One of the most common themes in contactee narratives was the warning against nuclear weapons. Contactees reported that extraterrestrials were concerned about the dangers posed by atomic weapons, not only to humanity but also to the stability of the broader cosmic order. According to these accounts, nuclear explosions disrupted not only Earth's atmosphere but also had consequences beyond our planet, potentially threatening other civilizations.

- **Extraterrestrials as Caretakers of Cosmic Balance**: Contactees like Adamski often described extraterrestrials as benevolent beings who acted as guardians of universal harmony. They warned that Earth's nuclear tests could destabilize the balance of cosmic forces, resulting in a "ripple effect" that affected other worlds. This narrative presented extraterrestrials as stewards of a larger interplanetary community, urging humanity to cease its destructive practices.
- **Fear of Contaminating Space**: Some contactees conveyed messages suggesting that nuclear fallout from atomic tests could reach beyond Earth, contaminating other celestial bodies or affecting the fabric of space itself. This concept echoed fears of environmental contamination on Earth, suggesting

that humanity's actions could have far-reaching consequences.

Emphasis on Spiritual Growth and Moral Evolution

In addition to warnings about nuclear weapons, extraterrestrials in contactee narratives often conveyed messages about the need for spiritual growth and moral evolution. They encouraged humanity to transcend its warlike tendencies and adopt values of compassion, unity, and cooperation. This theme resonated with a public grappling with the moral implications of nuclear warfare and the ideological divide of the Cold War.

- **The Cosmic Brotherhood**: Many contactees described the existence of a "cosmic brotherhood" or "galactic federation" composed of advanced, peaceful civilizations. According to these narratives, humanity was on the verge of joining this cosmic community but needed to overcome its divisions and aggression to be accepted. This vision of an interstellar alliance offered an idealistic alternative to the polarized reality of the Cold War.
- **A New Path for Humanity**: Contactees often presented extraterrestrials as spiritually advanced beings who had transcended materialism, greed, and violence. They urged humanity to follow a similar path, warning that failure to do so could result in catastrophic consequences. The extraterrestrial message became a moral call to action, urging humans to choose peace over war, compassion over conflict, and unity over division.

Science Fiction Meets Spirituality: The Contactee Movement's Cultural Appeal

The Contactee Movement was influenced by the popularity of science fiction, which exploded in the post-war years. The Space Age inspired the imagination of millions, with both fictional and speculative works depicting the possibility of contact with otherworldly civilizations. This era saw the rise of pulp magazines, novels, and films that explored space travel, alien life, and futuristic societies. These narratives provided a framework that allowed people to envision extraterrestrials not as terrifying monsters but as advanced, potentially benevolent beings.

Science Fiction as a Source of Hope

Science fiction offered a vision of the future in which humanity might overcome its limitations and make contact with other life forms. As space exploration moved from science fiction to reality, with the launch of Sputnik and the Apollo missions, people became more open to the idea that extraterrestrial life might not only exist but also interact with Earth. The Contactee Movement tapped into this sentiment, presenting extraterrestrials as cosmic neighbors with a vested interest in humanity's survival and growth.

- **The Space Brothers**: Contactees like Adamski introduced the idea of "Space Brothers," advanced beings from planets like Venus, Mars, and beyond, who were concerned with humanity's welfare. Unlike the hostile aliens of science fiction horror, the Space Brothers were friendly, enlightened beings who sought to guide humanity toward peace.
- **Blending Mysticism and Futurism**: The messages delivered by contactees combined elements of mysticism with futuristic ideals. The Space Brothers were described as beings who had achieved spiritual enlightenment, existing in harmony with the universe. This blend of spiritual and technological ideals appealed to individuals seeking alternatives to materialism and the prevailing ideologies of the Cold War.

The Countercultural Connection: The Contactee Movement's Influence on the 1960s and Beyond

The themes espoused by the Contactee Movement resonated with the emerging countercultural movements of the 1960s, which emphasized peace, spiritual exploration, and resistance to traditional authority. The contactees' warnings about nuclear war and calls for global unity found a receptive audience among those disillusioned with the mainstream values of Western society.

Aligning with Countercultural Ideals

The Contactee Movement's messages of peace, love, and unity echoed the ideals of the counterculture, which rejected materialism, embraced communal living, and advocated for social change. Many followers of the Contactee Movement were drawn to these ideals, seeing the extraterrestrials as champions of a more evolved way of life.

- **Environmentalism and Cosmic Responsibility**: Contactee narratives often included themes of environmental stewardship, echoing concerns that would later become central to the environmental movement. The Space Brothers' warnings about nuclear testing paralleled growing awareness of pollution and the need to protect the planet, linking cosmic consciousness with environmental responsibility.
- **Pioneers of the New Age Movement**: The Contactee Movement laid some of the foundations for the New Age movement, which gained prominence in the 1970s. Many contactees spoke of spiritual evolution, inner peace, and universal harmony, concepts that would later become central to New Age philosophy.

Skepticism and Criticism: Questioning the Legitimacy of Contactee Claims

While the Contactee Movement attracted followers, it was also met with skepticism. Many critics viewed the movement as a reaction to Cold War fears, suggesting that the contactees' narratives were shaped more by human anxieties than by genuine extraterrestrial encounters. Scientists and skeptics questioned the plausibility of beings from planets like Venus, pointing out that these environments were inhospitable to life as we know it.

- **Psychological Interpretation**: Some skeptics argued that the contactees' experiences were psychological in nature, possibly influenced by the collective trauma of the Cold War. The Space Brothers' warnings about nuclear war, they argued, reflected human fears and hopes rather than messages from alien beings.
- **The Role of Mass Media**: The rise of mass media played a significant role in spreading the Contactee Movement's narratives. Radio, television, and newspapers amplified the stories of the contactees, making them accessible to a broader audience. However, media coverage often portrayed contactees as eccentric or fraudulent, leading some to view the movement with suspicion.

A Movement Shaped by Fear, Hope, and the Quest for Meaning

The Contactee Movement was a unique product of its time, rooted in Cold War anxieties and inspired by the optimism of the Space Age. As humanity grappled with the possibility of nuclear annihilation, contactees emerged with messages of peace, unity, and spiritual evolution, delivered by beings who embodied humanity's idealized vision of cosmic wisdom. The movement captured the imaginations of thousands, blending science fiction, spirituality, and social idealism into a narrative that offered hope for a troubled world.

While the Contactee Movement eventually faded, its impact endures in the ongoing fascination with extraterrestrial life, the search for universal meaning, and the belief that humanity's fate may be intertwined with forces beyond our understanding. Whether viewed as a genuine phenomenon or a reflection of human aspiration, the Contactee Movement serves as a testament to humanity's resilience, creativity, and relentless quest for answers in the face of existential uncertainty.

From Friendly Encounters to Fearful Abductions: The Transition from Contactee Narratives to the Modern Abduction Phenomenon (1970s–1990s)

The 1970s saw a dramatic shift in how extraterrestrial encounters were perceived. Where once the Contactee Movement of the 1950s and early 1960s was marked by individuals who claimed to communicate with benevolent "Space Brothers," the modern abduction phenomenon introduced a new, unsettling theme. Individuals now reported being forcibly taken by extraterrestrials, often experiencing invasive medical procedures and intense psychological trauma. This shift from peaceful messages to fearful experiences mirrors larger cultural and psychological shifts in society, as well as evolving perceptions of alien life in popular culture.

Understanding the transition from the Contactee Movement to the abduction phenomenon requires examining the social context, examining the psychological factors at play, and exploring the impact of media and science fiction on public perceptions of extraterrestrial encounters.

The End of the Age of "Space Brothers": From Contact to Abduction

In the 1950s and early 1960s, the Contactee Movement portrayed extraterrestrials as enlightened beings. These "Space Brothers" communicated messages of peace, unity, and spiritual advancement, warning humanity of the dangers of nuclear weapons and environmental degradation. Contactees like George Adamski and Howard Menger described their encounters with extraterrestrials as positive, enlightening experiences, portraying these beings as mentors eager to guide humanity toward a better future. However, by the late 1960s and into the 1970s, this narrative began to change.

A Shift in Tone: Growing Distrust and Alienation

As the 1960s gave way to the 1970s, social and political unrest became more pronounced, and public trust in institutions began to erode. The Vietnam War, political scandals, and rising environmental and economic challenges made optimism about human progress increasingly difficult to sustain. These broader social anxieties found reflection in a more unsettling depiction of extraterrestrial encounters.

- **Loss of Innocence**: While the Space Brothers were seen as guides for humanity, the new narrative depicted extraterrestrials as indifferent or even malevolent. This shift reflects a loss of innocence, with encounters now framed as traumatizing rather than enlightening.
- **Growing Paranoia and Alienation**: The sense of isolation and suspicion in American society—fueled by events like Watergate and economic challenges—resonated with the themes of alien abduction. People began to feel a lack of control over their lives, an emotion mirrored in accounts of being helplessly taken by extraterrestrial forces.

First Reports of Alien Abduction: The Pioneering Cases of the 1960s

The seeds of the abduction phenomenon can be traced back to the 1960s with two significant cases: the abduction of Betty and Barney Hill in 1961 and the experience of Herbert Schirmer in 1967. These early cases set the stage for a flood of similar reports in the following decades.

Betty and Barney Hill: America's First Abduction Story

The abduction of Betty and Barney Hill is often regarded as the first widely reported alien abduction case in the United States. In September 1961, the Hills, an interracial couple from New Hampshire, reported seeing a strange light in the sky while driving home. Under hypnosis, they later recounted being taken aboard a spacecraft, where they underwent a series of medical examinations conducted by small, grey-skinned beings.

- **Unsettling Details and Medical Procedures**: The Hills described invasive medical procedures, including the extraction of skin and bodily fluids. Unlike the benevolent Space Brothers, these beings seemed clinical, detached, and indifferent to the Hills' feelings.
- **Cultural Impact**: The Hills' account marked a turning point, introducing themes of fear, helplessness, and violation that would become hallmarks of later abduction narratives. Their story, widely publicized, sparked a surge of similar reports and became foundational to the abduction phenomenon.

Herbert Schirmer's Encounter: Establishing the Grey Alien Archetype

In 1967, police officer Herbert Schirmer reported an encounter with grey-skinned extraterrestrials after seeing a

UFO while on patrol in Nebraska. Schirmer described the beings as humanoid but emotionally distant, and he claimed that they communicated telepathically. This account further cemented the image of the "Grey" alien, a depiction that would become dominant in abduction narratives.

The 1970s and 1980s: Alien Abduction Becomes a Cultural Phenomenon

By the 1970s and 1980s, the abduction phenomenon gained widespread attention, with hundreds of individuals reporting similar experiences. Certain themes became recurring, including paralysis, bright lights, abduction from bedrooms, and experiences of medical examination. The rise in abduction reports during this period paralleled an increasing interest in paranormal phenomena and a societal shift toward more skeptical and fearful views of alien life.

Patterns in Abduction Narratives: Common Experiences and Themes

Abduction accounts during this period often included highly specific details that became nearly universal in abduction narratives. These shared characteristics contributed to the formation of a "standard" abduction experience, with abductees describing similar events even when they had no prior knowledge of other cases.

- **The Bedroom Encounter**: Many abductees reported being taken from their beds at night, often experiencing a state of paralysis as bright lights filled their rooms. This setting created an atmosphere of vulnerability, as people described feeling invaded in their most personal space.
- **Invasive Medical Examinations**: The focus on physical examinations in abduction narratives reflects a shift from spiritual themes to clinical ones. Abductees often reported procedures involving needles,

scalpels, and other medical instruments, describing experiences that were painful and humiliating.
- **Memory Suppression and Hypnotic Recall**: Many abductees claimed to have "missing time" following their experiences, with memories that were fragmented or suppressed. Hypnosis became a common tool to retrieve these memories, often revealing vivid, emotionally charged details that left a lasting psychological impact.

The Grey Alien Becomes the Icon of Abduction

The image of the Grey alien, a small humanoid with large black eyes, became the dominant depiction in abduction accounts during this period. The cold, emotionless demeanor of the Greys and their unsettling physical appearance embodied the fear and mistrust associated with abduction experiences.

- **A Distinctly Alien Presence**: The Greys were described as lacking empathy, often treating abductees as experimental subjects rather than sentient beings. This image starkly contrasts with the warm, human-like Space Brothers of the Contactee Movement and may reflect an evolution in how society viewed extraterrestrial life.

Societal Influence on Abduction Narratives: Fear, Technology, and Trauma

The transition from contactee narratives to abduction experiences can be understood as a response to broader societal changes. The 1970s and 1980s were marked by significant shifts in technology, increased public interest in psychology, and an evolving understanding of trauma, all of which influenced how abduction stories were shaped and interpreted.

Technology and Medical Advancements: Echoes in Abduction Narratives

During this period, advances in medical technology and new forms of psychological therapy were reshaping public perceptions of the body and mind. The rise of clinical imagery in abduction accounts reflects these technological changes, as well as a growing fascination with (and fear of) medical procedures.

- **Medicalization of Extraterrestrial Encounters**: The emphasis on medical examinations in abduction narratives parallels a period of rapid medical advancements and heightened awareness of invasive procedures. As society became more familiar with medical technology, abductees began describing encounters that mirrored real-world medical experiences, but with an alien twist.
- **Fear of Loss of Autonomy**: The clinical detachment of the Greys and the focus on involuntary medical procedures reflect societal fears about the erosion of personal autonomy in an increasingly technological world. Abductees often described feeling like mere objects or subjects of experiments, a reflection of anxieties about being powerless in the face of advancing technology.

The Rise of Trauma Awareness: Shaping Abduction Narratives

In the 1970s and 1980s, the field of psychology began to develop a greater understanding of trauma, especially related to abuse and post-traumatic stress disorder (PTSD). Abduction narratives began to incorporate themes of trauma, fear, and helplessness, with abductees reporting symptoms that resembled PTSD, such as flashbacks, nightmares, and severe anxiety.

- **Psychological Impact and Memory**: Abductees frequently experienced memory suppression or "missing time," with hypnosis often used to recover details of their experiences. This process mirrored therapeutic techniques used to help trauma victims, reinforcing the perception that abduction experiences could have lasting psychological effects.
- **Trauma as an Interpretive Framework**: As awareness of trauma and its effects became more widespread, abductees and researchers alike began to interpret abduction experiences through this lens. Abductees described feelings of violation and powerlessness, viewing their encounters as traumatic events that left deep emotional scars.

The Role of Media and Popular Culture: Reinforcing the Abduction Narrative

The growing popularity of abduction stories in the media and science fiction played a critical role in shaping the abduction phenomenon. Television shows, movies, and books introduced extraterrestrials as enigmatic, often malevolent figures, reinforcing the themes of fear and helplessness that characterized abduction accounts.

Hollywood's Influence: Extraterrestrials as Threats

The portrayal of extraterrestrials in media shifted from friendly explorers to menacing figures with ambiguous intentions. Movies like *Close Encounters of the Third Kind* (1977) depicted aliens as powerful and inscrutable, while *The X-Files* (1993) popularized the idea of government cover-ups and abductions.

- **Popularizing the Grey Alien**: The Grey alien became a cultural icon in the 1980s, with its distinctive appearance featured in movies, television, and literature. This image reinforced the belief that

extraterrestrials were indifferent or hostile beings, contributing to a collective fear of alien encounters.
- **Media as a Reinforcing Agent**: The media's portrayal of alien abductions amplified the themes of helplessness and trauma, creating a feedback loop that validated abduction experiences. As more people encountered stories of abductions, the abduction narrative solidified, shaping how people interpreted their own unusual or unexplained experiences.

From Enlightenment to Fear—The Evolution of Extraterrestrial Encounters

The transition from contactee narratives to the abduction phenomenon represents a significant shift in the way extraterrestrial encounters were experienced and interpreted. While the Contactee Movement offered a hopeful vision of benevolent beings guiding humanity, the abduction phenomenon introduced themes of fear, trauma, and helplessness. This shift reflects broader social anxieties, technological advancements, and changing cultural narratives, as society moved from a period of optimism to an era marked by paranoia, alienation, and a loss of control.

The modern abduction phenomenon remains a powerful and enduring aspect of the extraterrestrial narrative. By understanding the factors that shaped this transition, we gain insight into the complex relationship between cultural fears, psychological trauma, and the ways in which humanity interprets encounters with the unknown. The evolution from contactee to abductee underscores the deep and shifting impact that extraterrestrial phenomena have had on society, reflecting our collective hopes, fears, and the ever-present quest to understand what lies beyond.

Warnings from Beyond: Common Themes in Abduction Messages (1970s–1990s)

The modern abduction phenomenon that emerged in the 1970s and continued into the 1990s introduced not only eerie stories of encounters with extraterrestrial beings but also powerful, recurring messages. Unlike the earlier Contactee Movement, where the narratives centered around friendly "Space Brothers" offering guidance, the messages reported during abductions carried an urgent tone, often warning of imminent crises on Earth. These messages touched on themes of environmental stewardship, human spirituality, and the need for global unity.

To understand these recurring messages within the context of the abduction phenomenon, it is essential to examine both the societal backdrop and the key themes that shaped these messages. From environmental warnings that foretold the impact of human recklessness to calls for spiritual evolution, these abduction messages reflect a complex tapestry of human hopes, fears, and moral dilemmas.

Contextual Shifts: An Era of Environmental and Spiritual Awakening

The 1970s marked a time of heightened environmental and spiritual consciousness in Western society. As pollution, industrialization, and the energy crisis became pressing concerns, the emerging environmental movement called for humanity to take responsibility for the planet's well-being. Meanwhile, interest in spiritual growth and consciousness exploration grew, with the rise of Eastern philosophies, New Age thought, and psychological self-awareness. These trends provided a fertile ground for the messages reported by abductees, as the themes of environmental and spiritual concerns resonated deeply with the anxieties and aspirations of the era.

- **Environmentalism on the Rise**: In 1970, the first Earth Day was celebrated, highlighting the growing awareness of environmental degradation. Issues such as pollution, deforestation, and wildlife extinction captured public attention, sparking demands for change.
- **Spiritual Exploration and the Human Potential Movement**: Interest in self-actualization and spiritual growth, promoted by figures like Carl Jung and Abraham Maslow, led many to explore consciousness beyond traditional religious beliefs. Concepts such as interconnectedness and global unity became central to New Age spirituality, a framework that lent itself well to the abduction messages about the need for spiritual evolution.

Environmental Messages: A Plea for Earth's Protection

One of the most prominent themes in abduction messages was the urgent need for environmental protection. Abductees reported that extraterrestrials expressed grave concerns about the damage humanity was inflicting on the planet. The messages warned of dire consequences if humans failed to take action to preserve Earth's ecosystems, suggesting that these extraterrestrial beings had a vested interest in the survival of the planet, if not humanity itself.

A Planet in Peril: Warnings of Ecological Catastrophe

Abductees described messages from extraterrestrial beings that warned of widespread environmental degradation, including pollution, deforestation, and climate change. These warnings often included detailed accounts of how human actions were affecting Earth's delicate balance, leading to species extinction, soil depletion, and rising temperatures.

- **Images of Destruction**: Some abductees reported seeing vivid images of environmental destruction, including landscapes ravaged by pollution, barren deserts where forests once stood, and seas choked with waste. These visions were presented as a cautionary glimpse into a potential future if humanity continued down its current path.
- **Focus on Pollution and Resource Depletion**: Abductees often received messages about the negative effects of industrial pollution and the unsustainable exploitation of natural resources. Extraterrestrials were depicted as deeply concerned with humanity's tendency to prioritize economic gain over environmental sustainability, suggesting that Earth's resources needed to be used more responsibly.

Protectors of the Earth: The Extraterrestrial Interest in Planetary Health

A recurring element in these environmental messages was the portrayal of extraterrestrials as stewards of the Earth, monitoring its health and occasionally intervening to prevent catastrophic outcomes. Abductees described these beings as deeply invested in Earth's ecosystems, with some claiming that extraterrestrials possessed knowledge of ancient natural processes and could foresee the long-term consequences of human actions.

- **Global Environmental Interconnectedness**: According to some abductees, extraterrestrials conveyed that Earth was part of a larger cosmic ecosystem, and that the planet's well-being was essential to the harmony of this greater system. This concept reinforced the idea that humanity's actions could affect more than just the immediate environment and even disrupt otherworldly realms.
- **Warnings of Climate Change and Catastrophic Events**: Abductees reported receiving warnings about climate change and other natural disasters that could

occur if humanity continued to ignore environmental concerns. While the term "climate change" was not commonly used at the time, these messages often referenced changing weather patterns, melting polar ice, and the risk of rising sea levels.

Spiritual Messages: Humanity's Need for Inner Evolution

Alongside environmental concerns, abductees frequently reported messages that addressed the spiritual growth and ethical evolution of humanity. These messages suggested that humanity was in a state of moral crisis and that a shift in consciousness was necessary to avert disaster. This focus on spiritual evolution echoed the ideals of the New Age movement and the human potential movement, both of which emphasized self-actualization and the transcendence of ego.

Calls for Compassion, Unity, and Higher Consciousness

Abductees described extraterrestrials urging humanity to develop a higher sense of compassion, unity, and empathy, suggesting that these traits were essential for survival. According to these messages, humanity's tendency toward division, materialism, and aggression was leading to self-destruction, and only a fundamental shift in consciousness could change this course.

- **Unity and Global Responsibility**: Extraterrestrials reportedly communicated that humans needed to overcome their divisions—be they political, racial, or cultural—to achieve a sense of unity. Abductees described messages that emphasized the interconnectedness of all beings, urging humanity to work together to solve global problems.

- **Warnings Against Materialism and Aggression**: Abductees often recounted that extraterrestrials criticized humanity's focus on material wealth and power. They warned that aggression, competition, and greed were traits that could lead to humanity's downfall, whereas qualities like empathy, humility, and cooperation were seen as vital for spiritual advancement.

The Evolution of Consciousness: Humanity's Role in a Cosmic Order

The messages also suggested that humanity had the potential for spiritual evolution and that this growth was necessary to join a "cosmic community." Abductees reported that extraterrestrials encouraged them to cultivate their inner awareness, develop a sense of cosmic responsibility, and align themselves with higher principles.

- **Integration into a Cosmic Brotherhood**: Many abduction messages included the idea of a cosmic brotherhood, an interstellar alliance of enlightened civilizations. According to abductees, extraterrestrials conveyed that humanity had the potential to join this alliance, but only if it could overcome its destructive tendencies and reach a higher level of consciousness.
- **Interconnectedness of Mind, Body, and Spirit**: Some abductees reported that extraterrestrials conveyed holistic messages about the interconnectedness of mind, body, and spirit. They emphasized the importance of meditation, self-reflection, and a harmonious relationship with nature as pathways to spiritual enlightenment.

A Duality of Warning and Hope: The Paradoxical Tone of Abduction Messages

The messages reported during abductions often presented a dualistic perspective: one of urgent warning, but also of hope. On one hand, extraterrestrials conveyed grave concerns about humanity's trajectory and the potential for irreversible damage. On the other hand, they seemed to offer a path forward, suggesting that if humanity could change its ways, it could avert disaster and join a higher cosmic order.

Warning as a Form of Guidance

Extraterrestrials' warnings about environmental and spiritual decline were often presented as a form of tough love, intended to wake humanity up to the consequences of its actions. These messages implied that extraterrestrials saw potential in humanity but believed that drastic change was needed.

- **Guidance for Avoiding Catastrophe**: Abductees reported that extraterrestrials provided specific guidance on how humanity could avert catastrophe, including promoting environmental stewardship, reducing reliance on harmful technologies, and fostering empathy and cooperation among nations.
- **Hope in Humanity's Capacity for Transformation**: Despite the warnings, these messages often conveyed a sense of hope that humanity was capable of transformation. Abductees described extraterrestrials as expressing optimism that, with the right guidance, humanity could evolve and take its place in a larger cosmic order.

The Role of Abductees as "Messengers" for Humanity

Abductees often felt they had been chosen by extraterrestrials to convey these messages to humanity, seeing themselves as intermediaries between Earth and the extraterrestrial realm. This sense of mission reinforced the belief that abduction experiences, while often traumatic, had a greater purpose.

- **Reluctant Prophets**: Many abductees reported that they initially resisted accepting the messages they received, feeling overwhelmed by the responsibility. However, over time, they came to see themselves as "reluctant prophets" tasked with sharing insights for the benefit of humanity.
- **Reinforcement of Abductees' Messages through Hypnosis**: Some abductees discovered these messages through hypnosis, which helped to retrieve suppressed memories of their encounters. Hypnotic regression played a significant role in reinforcing the belief that these messages were genuine, giving abductees the confidence to share them with a wider audience.

Extraterrestrial Messages as Reflections of Human Concerns

The environmental and spiritual messages reported during the modern abduction phenomenon from the 1970s to the 1990s reflect not only anxieties about the state of the world but also aspirations for a better future. These messages offered both warnings of potential disaster and a vision of hope, urging humanity to take responsibility for the planet and to develop a more compassionate, interconnected approach to life.

The dual themes of environmental preservation and spiritual evolution reveal a complex relationship between human concerns and the extraterrestrial narratives that emerged in this period. Whether interpreted as genuine messages from otherworldly beings or as psychological reflections of societal anxieties, these messages continue to resonate, inspiring those who seek to understand humanity's place within a larger, potentially cosmic order. The abduction phenomenon thus underscores humanity's enduring quest for meaning, guidance, and a future in which Earth and its inhabitants are part of a harmonious, universal community.

Humanity's Enduring Quest to Understand the Unknown

The exploration of extraterrestrial encounters and messages reveals a rich and evolving tapestry that stretches across centuries, weaving ancient myths, religious visions, Cold War fears, and modern abduction stories into a complex narrative of human experience. Each era has contributed unique interpretations of encounters with beings from beyond our world, influenced by the prevailing beliefs, anxieties, and aspirations of the time.

In ancient texts and myths, such as those found in the *Mahabharata* and the prophetic vision of Ezekiel, we see depictions of alien-like beings and celestial phenomena that resonate with later descriptions of extraterrestrial encounters. These early accounts blend the divine with the cosmic, reflecting humanity's intrinsic curiosity and desire to find meaning in the unknown. Similar accounts, like those of Plutarch's "flames in the sky" and Tacitus's sightings over Rome, suggest that unusual celestial events have long sparked wonder, awe, and interpretations that range from divine messages to harbingers of change.

Religious apparitions, including the Fatima Apparitions and the "Miracle of the Sun," provide a striking bridge between spiritual experience and modern UFO phenomena. As we cross-analyze religious messages with modern interpretations, a pattern emerges: messages often convey moral guidance, warnings, or calls for humanity to live in alignment with higher principles. This suggests that whether divine or extraterrestrial, these encounters often serve to reinforce or challenge humanity's ethical trajectory.

The Contactee Movement of the 1950s and 1970s marked a shift in the perception of extraterrestrial contact, presenting "Space Brothers" as benevolent guides concerned with humanity's welfare. Figures like George Adamski and Howard Menger shared messages of peace, spiritual growth, and environmental responsibility, set against a backdrop of Cold War tensions and nuclear fears. These narratives offered a hopeful vision of cosmic unity, suggesting that advanced civilizations had a vested interest in guiding humanity toward a peaceful future.

However, by the 1970s and 1990s, the tone of extraterrestrial encounters had shifted once more, with the rise of the abduction phenomenon. Unlike the friendly Space Brothers, abductors were often depicted as clinical, detached, and even invasive, particularly in abduction experiences reported under hypnosis. Yet, despite the fear and trauma associated with these accounts, abductees received recurring messages about environmental degradation and humanity's need for spiritual evolution. These messages echoed the era's growing awareness of ecological crises and the quest for spiritual enlightenment, reinforcing the sense that extraterrestrial contact might serve as a mirror, reflecting both humanity's greatest fears and highest aspirations.

In sum, the historical context of extraterrestrial encounters and messages underscores humanity's enduring fascination with the unknown and our longing for guidance, whether from gods, saints, or beings from distant stars. This journey through ancient texts, religious visions, contactee narratives, and abduction experiences reveals a pattern: each generation interprets these encounters through its unique cultural lens, shaping the messages to reflect contemporary concerns. Ultimately, these encounters remind us of our own capacity for wonder, our fear of the unknown, and our shared hope for a future in which humanity rises to meet its highest potential.

As we move from the historical context of extraterrestrial encounters into the core messages conveyed in these experiences, a new layer of understanding emerges. While the settings and details of these encounters have evolved, certain themes in the messages themselves remain remarkably consistent. From warnings about environmental degradation to cautionary advice on the dangers of nuclear weapons, these messages often reflect humanity's deepest concerns, echoing warnings that urge collective responsibility. Many of these communications, both subtle and overt, also emphasize a need for spiritual growth, suggesting that humanity's path forward requires not only technological advancements but also an evolution of consciousness. In this next section, we'll delve into specific types of messages reported in extraterrestrial encounters, analyzing how these communications have served as mirrors to our cultural anxieties and aspirations. Through case studies and cross-cultural analysis, we will explore whether these messages are cautionary, instructional, or prophetic, each one contributing to our complex relationship with the unknown.

Chapter 2: Guidance or Warning? Exploring the Core Messages in Extraterrestrial Encounters

The messages reported in extraterrestrial encounters often appear to transcend mere curiosity about humanity, addressing instead some of our most pressing challenges and moral dilemmas. Whether delivered through vivid imagery, telepathic impressions, or spoken words, these messages span an array of themes: urgent warnings about environmental destruction, cautions against the dangers of nuclear weapons, calls for spiritual evolution, and even prophetic visions of humanity's potential future. These communications seem to reflect both a recognition of humanity's potential and a concern for the consequences of our choices. In this section, we will examine the common themes of these extraterrestrial messages, analyzing the relevance of each in shaping our collective consciousness. Through specific case studies and cross-cultural comparisons, we'll uncover the underlying patterns that hint at either a guidance system from beyond or a reflective mirror of our own inner fears and hopes.

Warnings from Beyond: Extraterrestrial Messages on Environmental Destruction

As reports of extraterrestrial encounters have accumulated over the years, one of the most consistent and striking themes has been a warning about environmental destruction. Across diverse cases and geographical locations, many individuals who claim contact with extraterrestrial beings report receiving urgent messages that address humanity's impact on the planet. These communications often highlight critical issues such as climate change, pollution, and biodiversity loss, presenting a narrative that underscores the fragility of Earth's ecosystems and the need for immediate action.

To understand this aspect of extraterrestrial messages fully, it is essential to explore not only the content of these warnings but also how specific cases—like the Ariel School encounter, Travis Walton's experience, and even the Fatima Apparitions—reinforce a pattern. These cases reveal an emerging concern from alleged extraterrestrial sources about humanity's environmental impact, reflecting back some of the greatest fears of our time and serving as calls for collective responsibility.

Urgent Messages of Environmental Concern: A Common Theme

In many abduction and contact experiences, extraterrestrials reportedly convey dire messages about the state of Earth's environment. Witnesses often describe being shown vivid images of environmental degradation, with extraterrestrials warning of catastrophic consequences if humanity fails to change its behavior. These messages tend to address key areas of concern, from climate instability to pollution and the loss of biodiversity, aligning closely with modern environmental science.

Climate Change: A Planet in Peril

A recurring message in these encounters is the concern over climate change. Abductees and contactees alike report that extraterrestrials emphasize the consequences of unchecked carbon emissions, industrial activities, and the global rise in temperatures. Some experiences even include visions of extreme weather events, melting ice caps, and widespread natural disasters that suggest an imminent tipping point for the planet's climate systems.

- **A Global Responsibility**: Many individuals report that extraterrestrials communicated the need for humanity

to adopt a global perspective, recognizing that environmental harm in one region can have far-reaching effects. This message reflects an understanding of Earth as an interconnected system, where pollution and deforestation in one area contribute to climate instability worldwide.
- **Visions of a Future Earth**: In some encounters, extraterrestrials reportedly show abductees visions of Earth's future if humanity does not take action. These visions often include landscapes transformed by drought, flood-ravaged cities, and a scarcity of resources, all of which create a vivid and terrifying impression. By presenting these stark images, extraterrestrials seem to reinforce the urgency of the message, aiming to invoke a profound sense of responsibility in the witness.

Pollution: Humanity's Toxic Footprint

Another central theme in extraterrestrial warnings is the detrimental impact of pollution. These messages often highlight the effects of industrial waste, plastic, and chemical runoff on the planet's oceans, rivers, and air. Witnesses describe how extraterrestrials express grave concern over humanity's reliance on non-renewable resources and the unchecked disposal of toxic materials, which threaten the health of both ecosystems and humans.

- **Images of Contaminated Landscapes**: Some abductees report being shown images of polluted rivers, oceans teeming with plastic debris, and skies choked with smog. This imagery serves to illustrate the damage already done and the urgent need to reverse these trends.
- **The Warning Against Consumerism**: Linked closely with pollution is the theme of consumerism and wastefulness. Extraterrestrials reportedly convey messages about humanity's overconsumption and the unsustainable use of resources. Abductees frequently

describe messages suggesting that Earth's ecosystems cannot support such levels of consumption and that a shift toward sustainable living is essential for planetary survival.

Biodiversity Loss: The Irreplaceable Web of Life

In addition to concerns about climate and pollution, extraterrestrial messages frequently address the loss of biodiversity. Witnesses describe being told that humanity's activities are driving countless species to extinction, disrupting ecosystems, and threatening the delicate balance that sustains life on Earth. Extraterrestrials often present these warnings in stark terms, emphasizing that humanity's survival is intricately linked to the health of other species.

- **Depictions of Vanishing Species**: Abductees often describe being shown images of endangered animals or barren landscapes devoid of life. These visions highlight the magnitude of biodiversity loss, suggesting that extraterrestrials view the extinction crisis as a critical issue with potentially irreversible consequences.
- **The Importance of Ecosystem Balance**: Many extraterrestrial messages emphasize the interconnectedness of all life forms, suggesting that humanity's disregard for other species could lead to devastating consequences for Earth's ecological stability. This message echoes scientific understandings of ecosystem health, where the loss of biodiversity can weaken resilience and increase vulnerability to environmental change.

Case Studies: Extraterrestrial Messages on Environmental Destruction

Examining specific cases provides deeper insight into the environmental messages reported in extraterrestrial encounters. Each case offers a unique perspective on the content and urgency of these messages, from schoolchildren in Zimbabwe receiving warnings about Earth's future to individuals experiencing vivid, life-changing visions.

Ariel School Encounter (1994): Voices of Concern from the Youngest Witnesses

One of the most compelling cases of extraterrestrial warnings about environmental destruction took place in Ruwa, Zimbabwe, in 1994. Sixty-two schoolchildren at the Ariel School reported seeing a silver craft land nearby, after which small beings allegedly approached them. The children later recounted that these beings communicated telepathic messages about the future of Earth, emphasizing environmental concerns. Several children reported that the beings conveyed warnings about the harm humanity was causing to the planet and the urgent need for change.

- **The Impact on Young Minds**: The Ariel School case is particularly notable because of the age of the witnesses. Many of the children expressed that they had been shown images of destruction and desolation, including images of forests disappearing and natural resources vanishing. The children's accounts highlight the impressionability of young minds and suggest that extraterrestrial messages are aimed at reaching the next generation as much as the present one.
- **A Lasting Impression**: The Ariel School encounter left a profound impact on the witnesses, many of whom still recall the experience decades later. This case illustrates how environmental messages from extraterrestrials, particularly when communicated to children, have a lasting influence and reinforce a sense of environmental stewardship.

Travis Walton (1975): A Vision of Consequence and Responsibility

The case of Travis Walton, one of the most well-known abduction stories, also includes elements of environmental messaging. In 1975, Walton and his coworkers encountered a hovering UFO in the Apache-Sitgreaves National Forest in Arizona. Walton reportedly approached the craft and was subsequently taken aboard, where he claims to have experienced a series of vivid, dreamlike visions. While Walton's primary experience does not center on environmental warnings, he later described how the encounter gave him a heightened awareness of Earth's fragility and humanity's responsibility to protect it.

- **A Personal Awakening**: Although Walton's abduction narrative does not include explicit messages from extraterrestrials, his experience left him with a lasting impression of humanity's place in the natural world. Walton has since spoken about his sense of responsibility to protect the environment, suggesting that his experience indirectly inspired a deeper appreciation for Earth's ecosystems.
- **Indirect Environmental Themes**: The Travis Walton case demonstrates that not all extraterrestrial messages are overtly delivered. Some encounters, like Walton's, lead individuals to internalize broader environmental themes, which they later interpret as implicit messages about humanity's duty to care for the planet.

The Fatima Apparitions (1917): A Spiritual and Environmental Message?

While the Fatima Apparitions are primarily viewed as a religious phenomenon, some ufologists have reinterpreted this famous event as having extraterrestrial elements, particularly in light of the "Miracle of the Sun" witnessed by thousands. Some researchers suggest that the apparitions may have included an early warning about the future state of the planet, and the messages of repentance and moral correction align with later extraterrestrial messages about spiritual and environmental responsibility.

- **A Blending of Religious and Environmental Themes**: The Fatima messages included calls for repentance and a turning away from materialism, messages that resonate with the themes of environmental stewardship found in later extraterrestrial encounters. This blending of moral and environmental messaging suggests that similar themes have been present throughout history, albeit framed differently depending on cultural context.
- **The Miracle of the Sun**: During the Fatima event, witnesses described seeing the sun dance and spin in a way that defied natural laws. Some interpret this as a potential extraterrestrial phenomenon, suggesting that beings from beyond Earth might have been involved in delivering messages about humanity's moral and environmental responsibility.

The Underlying Message: Humanity's Role as Caretakers of Earth

The environmental warnings conveyed in extraterrestrial encounters present a unifying message: that humanity holds the responsibility for Earth's future. Extraterrestrials, as portrayed in these accounts, seem to recognize the interdependence of all life on Earth and emphasize that humans are guardians of a delicate and precious ecosystem. These messages imply that, without a shift in behavior and priorities, humanity risks damaging not only its own future but also the well-being of the planet and, by extension, the larger cosmic community.

Messages of Urgency and Hope

The messages related to environmental destruction reported in extraterrestrial encounters reflect both urgency and hope. On one hand, the warnings about climate change, pollution, and biodiversity loss underscore the immediate need for humanity to change its course. On the other, these messages suggest that it is not too late to make a difference. Cases like the Ariel School encounter, Travis Walton's experience, and even the reinterpretation of the Fatima Apparitions reveal a consistent narrative: that extraterrestrials view Earth as a valuable and interconnected ecosystem and urge humanity to protect it. Through these messages, humanity is presented with a choice to either continue down a path of environmental degradation or rise to the challenge of becoming responsible stewards of the planet, aligning with a vision of hope for a sustainable and interconnected future.

Warnings of Destruction: Extraterrestrial Messages on Nuclear Weapons and War

Among the most striking themes reported in extraterrestrial encounters is the urgent warning about humanity's development and deployment of nuclear weapons. As the world entered the nuclear age following the bombings of Hiroshima and Nagasaki, a new era of existential threat began—one that was not lost on alleged extraterrestrial visitors. Encounters involving extraterrestrial messages have often touched on the dangers of nuclear warfare, emphasizing the far-reaching and potentially catastrophic consequences of atomic weapons. Reports from military personnel, especially incidents at nuclear facilities such as Rendlesham Forest and Malmstrom Air Force Base, add a compelling layer of credibility to these warnings, suggesting that extraterrestrials may be particularly interested in humanity's nuclear capabilities.

To understand the full scope of these warnings, we must consider the historical context in which these messages emerged, the notable cases where these warnings were reportedly conveyed, and the ways in which these encounters continue to resonate with humanity's ongoing concerns about nuclear conflict and global peace.

The Dawn of the Nuclear Age: A New Era of Existential Threat

The detonation of the atomic bombs over Hiroshima and Nagasaki in 1945 not only brought an end to World War II but also marked the beginning of the nuclear era. This unprecedented destructive power transformed global politics, shaping a new world order and instigating an arms race that defined the Cold War. The threat of mutually assured destruction loomed large over humanity, making the specter of annihilation a defining feature of the 20th century.

Hiroshima and Nagasaki: A Turning Point for Humanity

The bombings of Hiroshima and Nagasaki served as a visceral demonstration of nuclear power. These events introduced the world to the horrifying reality of atomic warfare, with immediate and long-term effects that would forever alter global perceptions of warfare. The devastation caused by the bombings, along with the lasting impact of radiation, underscored the potential for nuclear weapons to threaten not only individual nations but the entire planet.

- **A Catalyst for Extraterrestrial Concern**: In the years that followed, reports of extraterrestrial encounters often mentioned the destructive potential of nuclear weapons. Many abductees and contactees claimed that extraterrestrials were concerned about humanity's capability to destroy itself and possibly even affect other realms or dimensions. These messages suggest that extraterrestrial beings, if they exist, may view

nuclear weapons as a profound threat to universal balance and stability.
- **The Atomic Age and the Fear of Annihilation**: The development of nuclear weapons brought with it the concept of mutually assured destruction, wherein both superpowers—the United States and the Soviet Union—maintained enough nuclear firepower to annihilate each other. This precarious balance of power created a pervasive sense of existential dread, influencing everything from public policy to popular culture. Extraterrestrial warnings about nuclear weapons seem to reflect this anxiety, urging humanity to reconsider the path it had taken.

The Cold War Context: An Era of Escalation and Fear

The Cold War between the United States and the Soviet Union escalated the threat of nuclear warfare to unprecedented levels. Both nations raced to build nuclear arsenals, leading to tensions and near-crises, such as the Cuban Missile Crisis of 1962, which brought the world to the brink of nuclear war. It was during this period that many of the most famous extraterrestrial encounters took place, with recurring themes of warnings about humanity's use of nuclear technology.

- **Extraterrestrials as Peacemakers?** Many contactees during the 1950s and 1960s reported that extraterrestrial beings communicated messages about the need for global unity and nuclear disarmament. These beings, sometimes referred to as "Space Brothers," were portrayed as wise and enlightened, advocating for peace and urging humanity to avoid the path of destruction.
- **The Fear of Global Catastrophe**: The possibility of nuclear war was a constant undercurrent in everyday life during the Cold War. Civil defense drills, fallout shelters, and government broadcasts reinforced the

idea that a nuclear strike could happen at any time. This pervasive fear is reflected in the warnings reported during extraterrestrial encounters, with many messages explicitly addressing the potential for global catastrophe if humanity failed to disarm.

Reports from Military Personnel: Warnings and Encounters at Nuclear Facilities

One of the most compelling aspects of extraterrestrial warnings about nuclear weapons comes from encounters reported by military personnel at nuclear installations. These incidents often involve sightings of unidentified flying objects (UFOs) near sensitive nuclear facilities, where these mysterious objects allegedly interfered with equipment and demonstrated a concerning level of control over nuclear weapons systems.

The Rendlesham Forest Incident (1980): A Close Encounter with Implications

The Rendlesham Forest incident, often referred to as the "British Roswell," took place in December 1980 near the Rendlesham Forest in Suffolk, England. U.S. Air Force personnel stationed at RAF Woodbridge reported a series of unusual events, including sightings of strange lights and encounters with a triangular craft. Some witnesses even claimed to have received telepathic messages during the encounter, which included a sense of warning about nuclear weapons.

- **Eyewitness Accounts from Military Personnel**: One of the most notable aspects of the Rendlesham Forest incident is that it involved highly trained military personnel, who provided detailed accounts of the sightings. These witnesses described seeing a craft that emitted bright lights and appeared to land in the forest. Some claimed to have experienced a strange

sense of calm and received mental impressions that hinted at extraterrestrial concern over humanity's destructive potential.
- **The Nuclear Connection**: While the Rendlesham Forest incident did not take place at a nuclear facility, the involvement of military personnel and the perceived warning about nuclear weapons add a significant layer to the narrative. The implication that extraterrestrial beings might be monitoring humanity's military activities, especially those involving nuclear capabilities, suggests that these entities view nuclear arms as a significant threat.

Malmstrom Air Force Base Incident (1967): Unidentified Objects and Disabled Missiles

Perhaps one of the most famous and well-documented cases of extraterrestrial involvement at a nuclear facility occurred in 1967 at Malmstrom Air Force Base in Montana. During this incident, multiple missile silos at the base reportedly went offline following the appearance of a UFO over the facility. This encounter involved the disabling of ten intercontinental ballistic missiles (ICBMs), rendering them temporarily inoperable.

- **Disabling of Nuclear Weapons Systems**: According to witnesses, a large, glowing object was seen hovering near the missile silos. Shortly thereafter, the missiles went offline, and all attempts to restore them failed temporarily. This incident raised alarms within the military, as it demonstrated a level of control over nuclear systems that defied explanation. Some witnesses later suggested that this was an intentional display of extraterrestrial disapproval of nuclear weapons.
- **Testimonies of Former Military Personnel**: Over the years, numerous former military personnel have come forward to corroborate the events at Malmstrom Air Force Base. These individuals claim that the incident

was covered up by the military, who were reluctant to acknowledge that unidentified objects had interfered with the U.S. nuclear arsenal. The implication that extraterrestrials could disable nuclear weapons has profound implications for national security and global stability, suggesting that these entities may be actively discouraging the use of atomic weaponry.

Themes in Extraterrestrial Warnings: Destruction, Unity, and Survival

The warnings about nuclear weapons conveyed during extraterrestrial encounters often focus on themes of destruction, unity, and survival. These messages, whether reported by military personnel or civilian contactees, reflect a deep concern over humanity's potential to self-annihilate. At the same time, they emphasize the need for unity, urging humanity to transcend nationalistic divisions and work together to ensure survival.

The Threat of Self-Annihilation: A Shared Concern

The primary theme in these warnings is the potential for nuclear weapons to lead to humanity's self-destruction. Extraterrestrials reportedly express concern not only for humanity but for the broader implications of nuclear warfare, which could disrupt ecosystems, harm the planet, and potentially impact other dimensions or realms of existence.

- **The Ripple Effect of Nuclear Weapons**: Some contactees report that extraterrestrials conveyed a belief that nuclear explosions could have far-reaching effects, impacting other dimensions or disturbing the cosmic order. This idea suggests that extraterrestrials view nuclear weapons as a threat that extends beyond Earth, underscoring the urgency of their warnings.

- **Humanity's Role in the Cosmos**: Many of these messages emphasize that humanity is part of a larger cosmic community and that its actions have implications beyond Earth. By warning about the dangers of nuclear weapons, extraterrestrials appear to be encouraging humanity to consider the broader consequences of its technological advancements, promoting a sense of interconnectedness and responsibility.

The Call for Unity: Transcending National Divisions

Alongside the warnings about nuclear weapons is a call for global unity. Extraterrestrials reportedly convey that humanity's divisions—political, national, and ideological—are a barrier to achieving lasting peace. These messages often suggest that only by working together can humanity avoid the devastation of nuclear war.

- **A Vision of Global Cooperation**: Many contactees describe messages that advocate for a world without borders, where humanity functions as a unified whole rather than competing factions. This vision of global cooperation aligns with the idea that humanity must transcend its divisions to ensure survival and prosperity.
- **Nuclear Disarmament as a Path to Peace**: Extraterrestrials often emphasize the importance of nuclear disarmament, suggesting that the elimination of nuclear weapons is a necessary step toward global harmony. This message resonates with the disarmament movements that gained traction during the Cold War, reinforcing the belief that peace can only be achieved through the renunciation of destructive technologies.

A Timeless Warning in a Nuclear Age

Extraterrestrial warnings about nuclear weapons and war highlight a recurring message that humanity has heard, in various forms, throughout the nuclear age. From the devastation of Hiroshima and Nagasaki to the near-misses of the Cold War, these encounters underscore the ever-present threat of nuclear conflict and the potential for humanity to shape its own fate. Through incidents at military facilities like Malmstrom Air Force Base and encounters like Rendlesham Forest, extraterrestrial warnings serve as a reminder of the stakes involved in humanity's pursuit of technological power.

The themes of destruction, unity, and survival resonate deeply, offering both a cautionary tale and a vision for a peaceful future. Whether one views these messages as real or symbolic, they speak to the profound responsibility humanity holds as stewards of a fragile world with the power to either preserve or destroy it. The warnings conveyed through these encounters serve not only as an urgent plea to reconsider the path of nuclear proliferation but as an invitation to transcend division and embrace unity in the shared quest for survival and peace.

A Path to Higher Consciousness: Extraterrestrial Messages of Spiritual Enlightenment and Evolution

Among the recurring themes in extraterrestrial encounters, messages of spiritual enlightenment and personal evolution stand out as a profound, often mystical component. These communications are typically relayed through telepathic impressions or visions, carrying insights and guidance on humanity's potential for growth. Unlike warnings about environmental and nuclear threats, which emphasize humanity's responsibilities, messages of spiritual enlightenment focus on inner transformation and the cultivation of higher states of consciousness. These experiences, often described by abductees and contactees, parallel many concepts found in religious teachings and New Age spirituality, suggesting that humanity's spiritual evolution may be central to its survival and its connection to the cosmos.

To fully explore these messages, it is essential to delve into how they are transmitted through telepathic encounters, what they reveal about humanity's capacity for spiritual growth, and how they resonate with long-standing religious and New Age concepts. The journey through these messages reveals a potential roadmap for human evolution, offering a vision of unity, compassion, and self-actualization.

Telepathic Encounters: Messages Beyond Words

Unlike spoken communication, many reports of extraterrestrial messages about spiritual enlightenment are conveyed through telepathic means. Contactees and abductees describe these experiences as profound, often emotional, and deeply personal. Telepathic communication in these encounters is reported to feel both immediate and immersive, carrying more than just words—it transmits understanding, emotions, and concepts that surpass linguistic boundaries.

A Language of the Mind: Telepathy as a Bridge to Spiritual Insight

Telepathic communication with extraterrestrials is described as an instantaneous transfer of ideas, where the receiver perceives not only thoughts but the essence of the message. This mode of communication transcends language barriers and cultural limitations, suggesting that extraterrestrials may view telepathy as the most effective way to convey complex, universal ideas about spiritual growth.

- **The Emotional and Sensory Depth of Telepathic Messages**: Witnesses often report feeling emotions or sensations accompanying the messages, as if the extraterrestrials are conveying not only intellectual information but an experiential understanding. This intensity creates a profound impact, leaving many contactees with a sense of urgency to act on the spiritual guidance they receive.
- **Messages of Inner Transformation**: Through telepathy, extraterrestrials reportedly urge humanity to look inward and cultivate qualities such as compassion, mindfulness, and interconnectedness. Contactees describe receiving guidance on practices for self-reflection and inner peace, often mirroring meditation techniques or spiritual exercises aimed at raising consciousness.

Universal Themes in Telepathic Messages

The content of these telepathic messages often revolves around themes that are universal and timeless, such as the need for inner harmony, unity with others, and alignment with higher values. Contactees frequently report that extraterrestrials emphasize the importance of developing qualities that transcend ego and cultivate a sense of collective responsibility.

- **Unity and Oneness**: Many messages center on the idea of unity, suggesting that humanity is part of a larger cosmic community. Extraterrestrials reportedly convey that humans need to overcome divisions and embrace a sense of interconnectedness, not only with each other but with all life. This emphasis on unity aligns closely with spiritual teachings that advocate for the dissolution of ego and the recognition of universal consciousness.
- **Compassion and Love as Pathways to Evolution**: Another recurring theme in telepathic messages is the call to cultivate compassion and love as essential steps

in spiritual evolution. Contactees describe extraterrestrials expressing concern over humanity's tendency toward conflict and competition, urging individuals to embrace empathy and cooperation as means to progress spiritually. This guidance mirrors spiritual traditions that emphasize love and compassion as fundamental to personal and collective growth.

Parallels with Religious and New Age Teachings: Ancient Wisdom and Modern Spirituality

The themes reported in these messages bear a striking resemblance to religious teachings and the principles of the New Age movement, which emphasize spiritual awakening, consciousness expansion, and a holistic view of humanity's place in the universe. This overlap suggests that the messages reported in extraterrestrial encounters may be tapping into universal archetypes and wisdom traditions, offering humanity insights that resonate across cultures and ages.

Ancient Teachings and the Path to Enlightenment

Many of the messages described by contactees align with the core teachings of major religious and spiritual traditions, from Eastern philosophies to mystical branches of Western religions. These messages often reflect the principles of enlightenment, compassion, and transcendence, all of which are central to spiritual growth in traditions like Buddhism, Hinduism, and Christianity.

- **The Concept of Self-Realization in Eastern Philosophy**: In Hinduism and Buddhism, enlightenment involves the dissolution of ego and realization of oneness with the universe. Contactees report similar guidance from extraterrestrials, who

convey that humanity must overcome selfishness and embrace interconnectedness as part of its spiritual evolution. This emphasis on self-realization and unity suggests that extraterrestrial messages echo ancient spiritual paths.
- **The Mystical Experience in Christianity**: Mystical Christianity emphasizes a personal connection with the divine and the cultivation of love and humility. The telepathic messages reported by contactees similarly advocate for love, compassion, and humility as pathways to spiritual enlightenment, reflecting the universal ideals found in mysticism. The sense of divine connection and guidance in these encounters mirrors the mystical experience of direct communication with a higher power.

New Age Concepts of Consciousness and Higher Beings

The messages reported in extraterrestrial encounters also align closely with New Age spirituality, which emerged in the 20th century as a fusion of Eastern philosophies, Western mysticism, and scientific theories about consciousness. The New Age movement advocates for a holistic approach to life, emphasizing interconnectedness, self-awareness, and the exploration of higher states of consciousness.

- **Extraterrestrials as Enlightened Beings or Ascended Masters**: In New Age thought, extraterrestrials are often seen as highly evolved entities who have attained wisdom through spiritual evolution. This interpretation views extraterrestrials as "ascended masters" who can guide humanity along a similar path of growth. Contactees report messages that suggest extraterrestrials are here to teach, to warn, and to offer a path toward higher awareness, reinforcing this New Age view.
- **The Evolution of Consciousness as a Cosmic Mandate**: New Age spirituality emphasizes the idea

that humanity is on a path of consciousness evolution, a theme echoed in extraterrestrial messages about spiritual growth. These messages suggest that the next stage of human evolution is not physical but mental and spiritual, requiring individuals to cultivate inner qualities that align with a higher cosmic order.

The Call for Inner Transformation: A Shift in Worldview

At the heart of these messages is a call for humanity to adopt a new worldview—one that values interconnectedness, respects the unity of all life, and sees personal growth as integral to collective evolution. This worldview aligns with the principles of holistic spirituality, encouraging individuals to see themselves not as isolated beings but as part of a vast, interconnected web.

- **Personal Growth as a Gateway to Collective Peace**: The extraterrestrial messages emphasize that individual spiritual progress is essential for global peace. By cultivating inner peace and compassion, humanity can overcome conflict, competition, and division. This perspective mirrors New Age teachings that emphasize the ripple effect of personal transformation on the larger world.
- **The Spiritual Evolution of Humanity as Cosmic Responsibility**: Extraterrestrials reportedly convey that humanity has a role within a broader cosmic community and that spiritual evolution is necessary to fulfill this role. This idea suggests that human development is not isolated but part of a larger cosmic framework, in which each being contributes to universal harmony.

The Spiritual Blueprint for Humanity's Evolution

The messages of spiritual enlightenment and evolution reported in extraterrestrial encounters present a vision of humanity's potential for growth. Through telepathic communication, extraterrestrials convey universal themes of unity, compassion, and higher consciousness, suggesting that humanity's future depends not only on technological progress but on inner transformation. These messages resonate with ancient religious teachings and New Age concepts, offering a timeless blueprint for spiritual growth.

By examining these messages, we gain insight into an extraterrestrial perspective that sees humanity not as a threat but as a species with profound potential. This guidance encourages us to cultivate compassion, foster interconnectedness, and strive for personal and collective enlightenment. Whether viewed as divine inspiration, psychological archetypes, or genuine extraterrestrial wisdom, these messages serve as a powerful reminder of the path humanity can choose: one that leads to a future shaped by empathy, awareness, and the realization of a higher purpose.

Glimpses of a Troubled Future: Apocalyptic and Prophetic Messages in Extraterrestrial Encounters

In many extraterrestrial encounters, one of the most unsettling themes is the warning of an impending apocalypse or drastic shift in humanity's future. These apocalyptic messages often come with vivid depictions of catastrophic events, environmental upheavals, or societal collapse, leaving witnesses with a sense of urgency and often a set of survival instructions. Unlike general messages of peace or spiritual growth, these communications suggest an unavoidable crisis on the horizon, conveying that humanity must act decisively to avert—or survive—a future catastrophe. Through a cross-cultural analysis, it becomes evident that these messages mirror long-standing themes of apocalypse found in religious texts, folklore, and cultural beliefs from around the world.

To develop a full understanding of these apocalyptic messages, it is essential to examine the content and implications of end-of-world scenarios and survival guidance offered in encounters, followed by a comparative look at how these themes resonate with apocalyptic narratives in various cultures.

End-of-World Scenarios: Messages of Catastrophe and Renewal

Abductees and contactees who report receiving apocalyptic messages often describe encounters that emphasize the fragility of human civilization and the potential for cataclysmic change. The messages range from warnings about natural disasters and environmental collapse to visions of global conflict or even cosmic upheaval. These messages convey a dual sense of urgency and inevitability, suggesting that while some events may be beyond humanity's control, there are ways to prepare and adapt.

Environmental and Cosmic Catastrophe: Visions of a Dying World

In many reported encounters, extraterrestrials communicate visions of Earth in crisis, often showing devastating environmental events that signal the degradation of the planet's natural systems. These visions align closely with modern environmental concerns, suggesting that humanity's continued exploitation of the environment could lead to irreversible consequences.

- **Scenes of Natural Disasters**: Witnesses often describe being shown images of earthquakes, floods, volcanic eruptions, and severe storms, painting a picture of a planet in turmoil. These catastrophic events are presented as a natural outcome of humanity's disregard for Earth's ecosystems, implying that climate change and environmental destruction are likely to lead to global disaster.

- **Warnings of Cosmic Events**: Some messages go beyond Earth, with extraterrestrials predicting cosmic events that could impact the planet, such as asteroid impacts or solar flares. These visions reflect the awareness that humanity is part of a larger cosmic order and suggest that existential threats can come from both within and beyond Earth's atmosphere.

Human Conflict and Societal Collapse: Warnings of a Fractured World

In addition to environmental calamities, apocalyptic messages from extraterrestrials often focus on humanity's social and political instability, predicting conflict, war, and the breakdown of society. These messages emphasize the dangers of nationalism, militarism, and ideological divides, portraying these elements as precursors to self-destruction.

- **Visions of Global Warfare**: Some witnesses report receiving messages warning of impending wars that could escalate to a global scale, potentially involving nuclear or biological weapons. These warnings reflect a concern that humanity's aggressive tendencies and failure to achieve unity could lead to catastrophic consequences.
- **Societal Breakdown and Chaos**: Extraterrestrial messages sometimes predict a gradual decline in social order, describing a world where resource scarcity, economic collapse, and widespread violence create a dystopian reality. These visions often portray society as fractured and struggling, with humanity's survival dependent on cooperation and resilience.

Survival Instructions: Guidance for a New World

With the images of destruction often come survival instructions, as extraterrestrials reportedly provide insights into how humanity can prepare for or survive the coming changes. These instructions vary from practical guidance on self-sufficiency to spiritual practices intended to foster resilience and adaptability.

- **Self-Sufficiency and Preparedness**: Many encounters involve messages about the importance of self-reliance, such as learning to grow food, conserve resources, and live in harmony with nature. Some witnesses report receiving detailed guidance on sustainable living practices, which they interpret as essential skills for surviving in a world undergoing environmental collapse.
- **Spiritual Resilience and Adaptation**: In addition to practical advice, extraterrestrial messages often emphasize the importance of spiritual and psychological preparedness. Witnesses are encouraged to develop inner strength, adaptability, and a sense of unity, which are seen as crucial for facing the challenges of a transformed world. This spiritual resilience is presented as a means to rise above fear and embrace change, reflecting a vision of humanity's potential for rebirth after crisis.

Cross-Cultural Analysis of Apocalyptic Themes: A Universal Fear of the End

The apocalyptic messages reported in extraterrestrial encounters resonate with narratives of the end times found across cultures and religions. From ancient texts to modern New Age beliefs, these stories share common themes: warnings of destruction, purification, and the potential for renewal. By examining these parallels, we gain insight into how extraterrestrial messages might reflect, amplify, or even shape humanity's collective fears and hopes about the future.

Apocalyptic Themes in Major Religious Traditions

Religious apocalyptic narratives often serve as moral cautionary tales, warning humanity of the consequences of failing to live in harmony with divine laws. These stories present a cycle of destruction and renewal, where a period of chaos is followed by a return to order, often through divine intervention or a moral reawakening.

- **Christianity: Revelations and the Final Judgment**: The Book of Revelation in the Christian Bible describes a series of catastrophic events that precede the second coming of Christ and the Final Judgment. In this narrative, humanity's sins lead to natural disasters, wars, and the collapse of earthly institutions. This vision of apocalypse emphasizes the idea of purification, where only the righteous survive to enter a renewed world.
- **Hinduism: The Cycles of Yugas and Kali Yuga's End**: In Hindu cosmology, time is divided into cycles called yugas, each representing a phase of moral and social decay. The current age, known as the Kali Yuga, is characterized by corruption, conflict, and loss of spiritual knowledge. Hindu texts predict that this era will culminate in destruction, followed by a new cycle of growth and renewal.
- **Indigenous Prophecies: Warnings from the Hopi and Maya**: Indigenous cultures have their own apocalyptic prophecies. The Hopi, for example, believe in a series of "worlds" that have been created and destroyed due to humanity's failure to live in harmony. They warn that the current world may also end if humanity does not reconnect with its spiritual roots. Similarly, interpretations of the Mayan calendar predict cycles of transformation, though modern scholars debate the apocalyptic interpretation.

Modern Apocalyptic Narratives: New Age and Utopian Ideals

In addition to traditional religious texts, modern apocalyptic themes are found in New Age beliefs and popular culture. These narratives often incorporate scientific knowledge about environmental degradation and cosmic events, blending them with spiritual concepts of ascension, renewal, and collective transformation.

- **The New Age "Shift of Ages"**: New Age spirituality often promotes the idea that humanity is on the brink of a transformative "shift," moving from a materialistic paradigm to a more spiritually enlightened one. This shift is seen as both an opportunity for growth and a challenge, as only those who embrace higher consciousness will thrive in the coming age. The warnings of extraterrestrials about apocalypse align with these New Age themes, suggesting that humanity must evolve spiritually to survive.
- **Doomsday Prophecies in Popular Culture**: Movies, books, and media have long explored apocalyptic scenarios, often emphasizing humanity's resilience and potential for rebirth. These narratives serve as modern myths, reflecting collective fears about technology, environmental collapse, and societal disintegration. Extraterrestrial messages about the end of the world reinforce these fears, yet also introduce a glimmer of hope for renewal, mirroring popular themes of survival and redemption.

Interpreting Apocalyptic Messages: Reflections on Humanity's Fate

The apocalyptic messages reported in extraterrestrial encounters raise profound questions about humanity's destiny, purpose, and place in the cosmos. Are these messages predictions, warnings, or reflections of human fears projected onto imagined beings from beyond? Many witnesses interpret these messages as a call to action, believing that extraterrestrials are invested in humanity's survival and moral evolution.

- **Warnings as Calls for Responsibility**: For some, apocalyptic messages from extraterrestrials represent a call for humanity to take responsibility for its future. These messages suggest that while certain cosmic or environmental events may be beyond human control, the social and moral choices humanity makes can influence the outcome.
- **Prophecies as Psychological Archetypes**: Others interpret these messages through a psychological lens, viewing them as archetypes embedded in the human psyche. The concept of apocalypse—whether caused by war, environmental destruction, or cosmic events—may reflect a universal fear of mortality, change, and transformation. Through extraterrestrial encounters, these fears are given shape and voice, allowing individuals to confront and process them.
- **The Role of Hope and Renewal**: Despite the dire warnings, many apocalyptic messages from extraterrestrials contain an element of hope, suggesting that humanity has the potential to transcend its destructive tendencies. This idea aligns with religious and cultural beliefs in renewal, where periods of crisis are followed by rebirth. Extraterrestrials, in this view, serve as guides or guardians, urging humanity to evolve and survive.

The Cosmic Cautionary Tale of Apocalypse

The apocalyptic and prophetic messages reported in extraterrestrial encounters present a powerful narrative of potential destruction and transformation. From visions of environmental collapse to warnings of cosmic events, these messages reflect humanity's greatest fears and its potential for resilience. By examining these themes in the context of cultural and religious apocalyptic narratives, we see how extraterrestrial encounters serve as both a cautionary tale and an invitation to embrace collective responsibility.

Whether regarded as genuine predictions, symbolic reflections, or psychological projections, these messages encourage humanity to confront the fragility of its civilization and the importance of ethical and spiritual resilience. They suggest that the choices humanity makes now will determine its future, offering both a warning and a path forward: that in the face of crisis, we can choose renewal, unity, and evolution over division and destruction. Through this lens, apocalyptic messages become not just visions of the end but a call to rise and endure, offering humanity a second chance to build a world that honors life, compassion, and interconnection.

The Multifaceted Messages of Extraterrestrial Encounters

The messages reported in extraterrestrial encounters offer a diverse tapestry of insights, warnings, and guidance that touch on humanity's most pressing issues. Across these narratives, we find recurring themes that highlight the challenges humanity faces and the paths it might take to overcome them. Warnings about environmental destruction emphasize the urgency of addressing climate change, pollution, and biodiversity loss. In encounters like those at the Ariel School and with Travis Walton, these messages reflect a shared concern for the planet's future and remind humanity of its role as a steward of Earth's ecosystems.

Similarly, messages addressing the dangers of nuclear weapons and war underscore the existential threat posed by technological power unchecked by ethical restraint. Cases like the Rendlesham Forest incident and Malmstrom Air Force Base underscore that extraterrestrials might view nuclear weapons as a critical danger not only to humanity but potentially to the cosmos itself. Through these encounters, witnesses often feel urged to pursue disarmament, global cooperation, and the cultivation of peace.

Messages about spiritual enlightenment and evolution invite humanity to consider inner growth as essential to collective well-being. These telepathic communications advocate for compassion, unity, and higher consciousness, echoing ideals found in religious and New Age teachings. Such guidance suggests that human evolution is not only physical but spiritual, requiring individuals to cultivate a deeper sense of interconnectedness.

Finally, apocalyptic or prophetic messages bring forth visions of cataclysmic events that threaten humanity's survival. These messages often parallel ancient and modern apocalyptic narratives, serving both as a warning and an invitation to prepare, adapt, and grow. The call for resilience, unity, and ethical renewal reminds humanity of its potential to survive and transform, even in the face of profound crisis.

Together, these messages form a comprehensive dialogue with humanity—urging responsibility, compassion, and a commitment to higher principles. Whether viewed as genuine extraterrestrial guidance or as reflections of human fears and aspirations, these encounters inspire a path toward a more sustainable, peaceful, and spiritually aware future. Through these messages, humanity is reminded of both its vulnerabilities and its potential, challenged to forge a future that respects life, honors the Earth, and embraces unity across all boundaries.

As we move from the messages conveyed in extraterrestrial encounters to the methods of communication used to deliver these messages, a fascinating dimension of these experiences unfolds. The way these communications are transmitted—whether through telepathic impressions, symbolic imagery, or physical contact—adds layers of complexity and mystery to the narratives. Each method of communication seems to be carefully chosen, conveying meaning beyond words and often leaving a lasting impact on the experiencer. Telepathy, with its ability to transcend language, offers an intimate and direct transfer of ideas, while symbolic and visual communication uses imagery to convey intricate concepts, sometimes rooted in the experiencer's own cultural background. Physical contact during abductions introduces sensory experiences that further blur the line between perception and reality, often requiring memory retrieval techniques such as hypnosis to fully understand. Exploring these varied methods allows us to gain a deeper insight into how extraterrestrials—or the phenomenon itself—bridge the gap between their world and ours, shaping the way we receive and interpret these profound messages.

Chapter 3: Bridging Worlds: Exploring Methods of Communication in Extraterrestrial Encounters

The nature of communication in extraterrestrial encounters is as varied and intriguing as the messages themselves, offering a glimpse into the ways these beings—or the phenomenon they represent—bridge the gap between our worlds. Unlike traditional forms of human communication, these encounters often involve methods that bypass spoken language, challenging our understanding of perception, memory, and consciousness. Telepathic communication provides an immediate, unfiltered transfer of thoughts and emotions, creating intimate connections that transcend words. Symbolic and visual communication introduces rich layers of meaning through images, symbols, and even star maps, which often carry personal or cultural significance for the experiencer. Physical contact during abductions, on the other hand, engages the senses directly, leaving impressions that are sometimes retrieved only through hypnosis or regression techniques. In this section, we will delve into these unique methods of communication, examining the psychological implications, interpretive challenges, and the profound impact they have on those who experience them. Together, these methods reveal a fascinating dimension of extraterrestrial encounters, offering us clues about how information might be shared across the boundaries of human understanding.

Unspoken Connections: Telepathic Communication in Extraterrestrial Encounters

One of the most compelling and enigmatic aspects of extraterrestrial encounters is the phenomenon of telepathic communication. In numerous reports of extraterrestrial encounters, telepathic communication stands out as a unique and compelling method through which messages are conveyed. Unlike spoken language, which requires shared vocabulary, grammar, and even cultural reference points, telepathy bypasses these barriers, delivering messages directly into the mind of the experiencer. Reports of telepathic communication in extraterrestrial encounters often include detailed accounts of intense, vivid exchanges that seem to convey not only words but complex ideas, emotions, and even visual impressions in a matter of seconds. Telepathy in this context raises profound questions about consciousness, perception, and the boundaries of human psychology, presenting a unique opportunity to explore how communication might transcend the physical senses.

To fully grasp the implications of telepathic communication in these encounters, it is essential to examine the ways in which telepathy functions as a bridge between human and extraterrestrial consciousness, how individuals experience these telepathic messages, and the psychological implications of receiving information in such a direct and unfiltered way.

The Role of Telepathy in Extraterrestrial Messaging: A Language Beyond Words

In accounts of extraterrestrial encounters, telepathy is often described as the primary means of communication. Witnesses report that extraterrestrials transmit thoughts, emotions, or even entire ideas directly into their minds, often creating a feeling of complete immersion in the message. This method of communication is particularly effective in overcoming language and cultural barriers, as it enables the experiencer to "understand" the message on an intuitive level rather than relying on words alone.

A Universal Mode of Communication: Telepathy as a Bridge Across Species

Telepathic messages in extraterrestrial encounters are often described as clear and unambiguous, conveying a level of detail that spoken language might not capture. This phenomenon suggests that extraterrestrials may view telepathy as a universal form of communication, one that is unbounded by cultural and linguistic limitations. Witnesses describe these exchanges as feeling inherently truthful, sometimes with a sense of wisdom or authority that enhances the message's impact.

- **Instantaneous Understanding**: One of the most significant features of telepathic communication is the sense of instantaneous understanding. Witnesses often report that, in a single flash of insight, they receive entire concepts or ideas that would take several minutes to communicate verbally. This direct and immediate transmission conveys not only information but the emotional weight or urgency behind it, making the experience feel personal and deeply impactful.
- **Transference of Emotions and Imagery**: Telepathy in extraterrestrial encounters often includes an emotional or sensory dimension, where experiencers receive not only words but feelings and images. Witnesses describe sensing extraterrestrials' emotions—whether concern, urgency, or compassion—along with vivid mental images that accompany the message. This sensory quality adds layers of meaning, making the experience far more profound than mere verbal communication.

Experiencers as "Receivers": Telepathy as a Tool for Shaping Perception

Those who receive telepathic messages during encounters often describe the experience as passive rather than active, likening it to "receiving" rather than "thinking." In this state, the witness feels as though their mind is open to information from an external source, with no sense of control over the content or timing of the message. This process has led some researchers to suggest that telepathy allows extraterrestrials to directly shape or influence human perception, presenting thoughts in a way that feels authentic and fully integrated.

- **Altered Perception and Heightened Awareness**: Experiencers often report a heightened state of awareness during telepathic exchanges, as if their minds have expanded to accommodate the incoming message. This altered state may involve a feeling of detachment from ordinary thoughts or even an impression of "shared consciousness," where the experiencer momentarily perceives the world from the extraterrestrial's perspective. This shift can create a lasting impact, often leading witnesses to reevaluate their beliefs and assumptions.
- **A Sense of Infallibility**: Telepathic messages in extraterrestrial encounters frequently create a powerful impression of truth or authenticity. Witnesses often describe these messages as "absolute" or "undeniable," feeling that the telepathic mode of communication lacks the ambiguity or manipulation that words sometimes carry. This sense of certainty can make the messages highly influential, leaving a lasting impact on the experiencer's worldview.

Psychological Analysis of Telepathic Experiences: The Mind's Response to Nonverbal Communication

Telepathic communication presents a fascinating challenge for psychology, as it raises questions about perception, memory, and the mind's openness to external influence. While telepathy remains a largely unexplored area within mainstream psychology, analyzing the experiences reported in extraterrestrial encounters can provide insight into how the human mind might interpret nonverbal messages and the psychological mechanisms at work during these exchanges.

The Role of Cognitive Models: Making Sense of the Unseen

To understand telepathic experiences, researchers often apply cognitive models that explore how the brain processes and interprets information without external sensory input. Telepathic messages, by bypassing traditional sensory channels, may stimulate unique areas of the brain, prompting responses that are both psychological and physiological.

- **Pattern Recognition and Mental Imagery**: Cognitive psychology suggests that the brain's pattern recognition abilities play a crucial role in telepathic experiences. When receiving a telepathic message, the mind may rely on stored images, emotions, or ideas to "construct" the message in a coherent way. This process helps the experiencer make sense of abstract information, allowing them to interpret the message within a familiar framework of memories, symbols, and emotions.
- **The Power of Suggestion and Expectancy**: Some psychologists argue that the power of suggestion and expectancy may shape telepathic experiences, as individuals who anticipate contact with

extraterrestrials may subconsciously prepare for nonverbal communication. This expectancy could heighten sensory perception, create vivid mental images, and amplify emotions, leading witnesses to interpret these experiences as telepathic communication from an external source.

The Psychological Impact of Telepathic Encounters: Profound, Lasting Effects

The psychological impact of telepathic communication is often profound, leaving a lasting impression that goes beyond the immediate encounter. Many experiencers report that telepathic messages shift their perspectives, attitudes, and even life goals, suggesting that these encounters affect the psyche at a deep level.

- **A Shift in Identity and Worldview**: Experiencers of telepathic communication often undergo significant shifts in their worldview, feeling as though they have gained insight into universal truths or cosmic wisdom. This transformation may lead to lasting changes in personality, values, and spiritual beliefs, as experiencers integrate the messages they received into their lives.
- **Post-Telepathic Stress and Integration**: The intensity of telepathic messages can create psychological challenges, as experiencers grapple with information or emotions that feel overwhelming. Some witnesses report experiencing post-telepathic stress, struggling to reconcile their new perspectives with their former beliefs. Integration of these experiences may involve extensive self-reflection, support from others, or even therapeutic intervention.

The Mystery of Telepathy: A Glimpse into Shared Consciousness?

Telepathy in extraterrestrial encounters raises compelling questions about consciousness, perception, and the boundaries of the human mind. The directness of telepathic communication, combined with its emotional and sensory depth, suggests that extraterrestrials may perceive consciousness differently from humans, viewing it as a shared rather than isolated phenomenon. This idea aligns with theories of collective consciousness and raises intriguing possibilities about the nature of mind and thought.

- **Telepathy as a Universal Language**: The universality of telepathy suggests that consciousness might be a shared field, accessible to beings from various realms or dimensions. If telepathic communication is possible across species, it implies that thoughts and emotions may transcend physical limitations, pointing to a form of connection that is both personal and universal.
- **Implications for Human Potential**: The phenomenon of telepathic communication also invites exploration of human potential. If extraterrestrials can communicate telepathically, it suggests that humans might possess similar latent abilities, with the capacity to expand their awareness and connect directly with others. This possibility could inspire new avenues of research into consciousness and nonverbal communication, challenging traditional views of the mind as a self-contained entity.

Telepathy as a Window into a Larger Reality

Telepathic communication in extraterrestrial encounters presents a captivating glimpse into a form of interaction that transcends language, culture, and even sensory perception. Through telepathy, experiencers receive messages that carry an immediacy and authenticity that often surpass traditional forms of communication, leading to powerful emotional and psychological shifts. These experiences suggest that telepathy serves as a bridge between human and extraterrestrial consciousness, opening the mind to new levels of understanding and insight.

From a psychological perspective, telepathy challenges established theories of perception and consciousness, suggesting that the mind is capable of receiving information without direct sensory input. This phenomenon may reflect cognitive mechanisms of pattern recognition and expectation or could point to a deeper, more mysterious aspect of consciousness—one that connects all beings within a shared mental or spiritual field. Telepathic encounters thus invite humanity to explore the limits of perception, the nature of thought, and the potential for communication that defies physical boundaries. Whether seen as genuine inter-species dialogue or an unexplored function of the mind, telepathy in extraterrestrial encounters represents a profound and transformative experience that expands our understanding of what it means to truly connect.

Beyond Words: The Role of Symbolic and Visual Communication in Extraterrestrial Encounters

One of the most captivating aspects of reported extraterrestrial encounters is the use of symbolic and visual communication. Unlike spoken language, which relies on words and grammar, symbolic and visual communication uses images, symbols, and sometimes star maps to convey messages. This form of communication adds depth and mystery to these encounters, allowing experiencers to perceive messages on multiple levels, yet it also introduces challenges in interpretation. When symbols and images are involved, understanding the message requires more than language proficiency; it demands intuition, cultural awareness, and sometimes a knowledge of astronomy or spiritual iconography.

To gain a comprehensive understanding of this topic, we must explore the types of symbols and images commonly reported in encounters, the role of cultural background in interpretation, and the unique challenges that arise when translating visual language. Examining these components reveals not only how extraterrestrials might communicate but also the profound ways that symbols can shape human perception, memory, and meaning.

Visual Messaging: The Power of Images, Symbols, and Star Maps

In many reports of extraterrestrial encounters, witnesses describe receiving messages not through words or telepathic thoughts but through powerful images and symbols. These visuals often carry intricate meaning, representing complex concepts that might be difficult to express verbally. By presenting messages visually, extraterrestrials appear to bypass language altogether, using imagery to communicate universal ideas that transcend linguistic and cultural divides. However, these symbols are often open to interpretation, and their meaning can vary based on the experiencer's background and personal beliefs.

Symbols as Universal Language: Bridging Human and Extraterrestrial Perception

Symbols have long served as a universal language in human history, used in religious rituals, art, and even dreams. In the context of extraterrestrial encounters, symbols appear to play a similar role, conveying ideas and feelings in a way that feels both direct and ambiguous. Witnesses often report seeing geometric shapes, star maps, or even ancient symbols that resonate on an intuitive level, suggesting that extraterrestrials may use symbols to communicate ideas that reach beyond the limits of words.

- **Geometric Shapes and Sacred Geometry**: Many experiencers describe seeing shapes like circles, triangles, and pyramids during encounters, sometimes arranged in patterns or sequences that evoke a sense of order or harmony. These shapes are reminiscent of "sacred geometry" found in ancient cultures, where geometric forms represent cosmic order or spiritual truths. The use of such symbols may indicate that extraterrestrials are attempting to convey concepts about universal structure, balance, or the interconnectedness of all things.
- **Celestial and Cosmic Imagery**: Star maps and constellations are also commonly reported, suggesting that extraterrestrials may use these visuals to communicate information about their origin or location in the cosmos. Some witnesses describe seeing detailed star patterns or receiving an "impression" of a specific place in the universe, as if being given directions or a sense of orientation. This form of imagery reinforces the idea that extraterrestrials may want humans to understand their place in the larger cosmic network.
- **Archetypal and Mythic Symbols**: In some cases, the symbols reported resemble archetypal or mythological imagery, such as winged beings, animals, or even symbols reminiscent of ancient religious iconography. These symbols carry deep psychological resonance, invoking themes of guidance, protection, or enlightenment. By using archetypes that hold universal meaning, extraterrestrials may be tapping into the collective unconscious, seeking to convey messages that resonate on a subconscious level.

Star Maps: Clues to Origins and Intentions

Star maps hold a particularly fascinating role in extraterrestrial encounters. Some experiencers report seeing intricate star charts during their encounters, often displayed as holographic images or visions in their minds. These star maps are typically interpreted as representations of the extraterrestrials' place of origin or perhaps significant cosmic landmarks. In some well-known cases, experiencers have even been able to identify real star systems based on the images they saw, suggesting a possible intent to share concrete information about extraterrestrial origins.

- **The Case of the Zeta Reticuli Star Map**: One of the most famous instances of star maps in extraterrestrial encounters is the account of Betty Hill, who claimed to have seen a star map during her abduction experience in the 1960s. Hill later drew the map from memory, and it was eventually linked to the Zeta Reticuli star system by astronomers and researchers. This association led some to speculate that the extraterrestrials were attempting to communicate their place of origin. This case illustrates how star maps in encounters can serve as "cosmic signatures," possibly intended to establish a sense of credibility or transparency.
- **Navigational Maps and Galactic Coordinates**: Other experiencers report seeing star maps that seem to suggest a navigational purpose, as if the extraterrestrials are guiding them to a specific location or conveying information about interstellar travel. These maps sometimes feature coordinates, lines, or paths between stars, hinting at the extraterrestrials' understanding of space and their ability to traverse cosmic distances. For witnesses, these maps often feel like a glimpse into advanced technology or knowledge, igniting curiosity about the potential for intergalactic travel.

Decoding the Language of Symbols: Challenges in Interpretation

While symbols have the potential to convey complex messages, they also pose significant challenges in interpretation. Symbols are inherently ambiguous, meaning that their significance can vary widely based on personal, cultural, and even psychological factors. Decoding these messages requires a balance of intuition and analysis, as well as an awareness of one's own cultural biases and expectations. This process highlights the complexities of cross-species communication, where both sender and receiver must work within different frames of reference.

The Role of Cultural Background in Shaping Meaning

An individual's cultural background plays a crucial role in interpreting symbolic messages, as symbols that hold profound meaning in one culture may appear neutral or even confusing in another. In extraterrestrial encounters, witnesses often interpret symbols through the lens of their own beliefs, sometimes attributing religious or spiritual significance to what might be intended as simple information. This tendency to interpret symbols based on personal or cultural frameworks suggests that understanding these messages requires an awareness of cultural influences.

- **Religious and Spiritual Associations**: Many symbols reported in encounters have connections to religious or spiritual traditions, leading witnesses to interpret them as divine or supernatural messages. For example, geometric shapes like the circle or cross have spiritual meanings in multiple cultures, representing unity, wholeness, or transcendence. Witnesses who associate these shapes with their own faith may view the extraterrestrial messages as signs of divine guidance, even though the true intent may be more scientific or secular.
- **Archetypal Resonance and the Collective Unconscious**: Psychologists like Carl Jung have argued that symbols tap into the "collective unconscious," a shared reservoir of archetypes that resonate across cultures. When witnesses see symbols that evoke universal themes—such as a tree, a sun, or an eye—they may feel an instinctual connection to the

message, interpreting it as something inherently meaningful. This archetypal resonance allows the symbols to transcend cultural boundaries to some extent, enabling the message to be understood on a deeper, often subconscious level.

The Ambiguity of Symbolic Communication: Balancing Intuition and Analysis

The ambiguity of symbols means that interpretation is rarely straightforward, requiring experiencers to balance intuition with careful analysis. Symbols can have multiple meanings or associations, making it difficult to determine the sender's true intent. This challenge is compounded by the fact that witnesses often feel as though the symbols "belong" to them in some way, making objective interpretation more challenging.

- **Subjectivity and Personal Symbolism**: Each individual has a unique set of personal symbols—shapes, images, or colors that carry specific meanings based on personal experiences. When witnesses see symbols in extraterrestrial encounters, they may unconsciously project their own meanings onto the images, interpreting them through a personal lens. While this approach allows for deeper personal connection, it can also obscure the intended meaning, creating interpretations that are highly subjective.
- **The Difficulty of Cross-Species Interpretation**: Extraterrestrial symbols may carry meanings that are entirely unfamiliar to human experience, making them challenging or even impossible to interpret accurately. Without a shared cultural or perceptual framework, humans may struggle to grasp the full significance of these symbols, leaving room for misinterpretation or incomplete understanding. This difficulty raises questions about the limitations of symbolic communication and the potential need for more direct methods of interaction.

A Window into the Extraterrestrial Mind: The Significance of Visual Communication

The use of symbols and visuals in extraterrestrial encounters suggests that these beings may perceive communication in fundamentally different ways from humans. Where humans rely on language to express detailed ideas, extraterrestrials may view symbols as a more direct or efficient method of conveying information. This reliance on imagery raises intriguing questions about extraterrestrial cognition, suggesting that their minds may process and communicate information on a multidimensional level that integrates thoughts, emotions, and symbols simultaneously.

- **Visuals as a Multi-Layered Language**: The complexity of symbolic messages in extraterrestrial encounters hints at a communication system that operates on multiple levels, potentially allowing extraterrestrials to convey concepts that involve both logical and emotional components. This multi-layered language would be far richer than human speech, capable of transmitting ideas that resonate on intellectual, psychological, and spiritual levels all at once.
- **A Reflection of Extraterrestrial Thought Processes**: The use of symbols may reflect an extraterrestrial way of thinking that prioritizes patterns, relationships, and holistic understanding. Where human language breaks ideas down into discrete words, symbolic communication presents them as interconnected wholes, creating a more intuitive and instantaneous understanding. This suggests that extraterrestrials might possess cognitive abilities that emphasize synthesis over analysis, viewing reality as a web of interconnected ideas and energies.

Symbols, Stars, and the Art of Cross-Species Communication

Symbolic and visual communication in extraterrestrial encounters offers a compelling and multifaceted form of interaction that transcends the boundaries of language. Through symbols, images, and star maps, extraterrestrials appear to communicate complex ideas that reach beyond spoken words, engaging the experiencer's mind on intellectual, emotional, and even spiritual levels. This method of communication allows for profound resonance but also introduces challenges in interpretation, as symbols are open to personal and cultural influences.

Experiencers often struggle to decode these messages, balancing intuition with analysis to derive meaning from symbols that feel both familiar and alien. Cultural background, personal beliefs, and psychological factors all play a role in shaping the interpretation of these symbols, highlighting the complexities of cross-species communication. Nevertheless, these encounters offer a glimpse into a potential "language" of the cosmos, one that reflects the interconnectedness of all beings and challenges humanity to expand its understanding of communication.

Ultimately, symbolic and visual communication in extraterrestrial encounters invites us to see the universe in a new light, one where images hold as much meaning as words and where understanding arises from shared resonance rather than literal translation. Whether viewed as glimpses into alien minds or reflections of the human psyche, these symbols encourage us to explore the deeper dimensions of consciousness, bridging the gap between known and unknown in our quest for understanding.

Beyond Words: Exploring Physical Contact and Communication through Abductions

Extraterrestrial encounters that involve physical contact and abductions bring a deeply visceral dimension to the phenomenon of communication. Unlike telepathic messages or symbolic imagery, these encounters engage the body and senses directly, leaving experiencers with vivid, sometimes traumatic, memories. Physical contact often includes intense sensory experiences, sometimes creating lasting impressions or even physical marks that witnesses cannot easily explain. Abduction encounters frequently involve physical examinations or procedures that experiencers interpret as attempts at understanding human biology, further blurring the line between communication and experimentation.

The process of remembering these encounters can be complex, as many experiencers report partial or fragmented memories that resurface through hypnosis or regression techniques. This use of memory retrieval methods adds another layer to the phenomenon, raising questions about the validity of these memories and the reliability of methods used to access them. Understanding physical contact and communication through abductions requires a close examination of both the sensory experiences during these events and the psychological mechanisms involved in memory retrieval.

Physical and Sensory Experiences in Abductions: Communication Beyond the Mind

Abduction encounters are often reported as a blend of physical and sensory experiences that go beyond verbal communication, engaging the experiencer on multiple levels. Witnesses frequently describe sensations that are unfamiliar or heightened, creating an impression of complete immersion. These sensory experiences can range from feelings of weightlessness or paralysis to sensations of touch, light, and sound, often overwhelming the senses and creating intense, lasting memories. For many, the tactile nature of these encounters makes them feel exceptionally real, leaving them with the undeniable sense that something extraordinary has occurred.

The Power of Touch and Tactile Interaction: Feeling the Unknown

One of the most reported aspects of physical contact during abduction experiences is touch. Witnesses describe being touched by the extraterrestrial beings, sometimes in gentle ways and at other times in ways that feel invasive or clinical. This form of contact leaves a strong impression, as touch is one of the most direct and intimate forms of communication. For many experiencers, the memory of this tactile interaction is both haunting and inescapable.

- **Sensations of Restraint and Immobility**: A recurring feature in abduction experiences is the sensation of being restrained or immobilized, as if held down by an unseen force. This feeling of paralysis can be both physical and psychological, with witnesses describing an inability to move or resist. While unsettling, this experience appears to be a means of controlling the experiencer's body during physical examinations or procedures, creating an atmosphere of vulnerability.
- **Gentle and Clinical Touches**: Some witnesses describe the touch of extraterrestrials as unexpectedly gentle, feeling more like a careful examination than an act of aggression. In these cases, witnesses report feeling as though the beings are studying their anatomy with precision, sometimes tracing fingers along the body or applying gentle pressure to specific areas. This kind of touch suggests a scientific curiosity on the part of the extraterrestrials, who may be attempting to gather data or communicate their interest in human physiology.

Sights, Sounds, and Sensory Overload: Immersive Communication through Sensory Input

In addition to touch, many abduction experiences are accompanied by heightened visual and auditory sensations that contribute to the sense of immersion. Witnesses often describe rooms filled with bright, unnatural lights, humming or buzzing sounds, and even subtle vibrations. These environmental details create a sensory landscape that is alien in every sense, reinforcing the feeling of being in an entirely different realm.

- **Bright Lights and Sterile Surroundings**: Abduction encounters often take place in environments that witnesses describe as starkly lit and sterile, evoking imagery reminiscent of medical examination rooms. The lights are frequently described as intense, even blinding, creating an atmosphere that feels clinical and detached. This setting enhances the feeling of being examined or studied, conveying a message of scientific detachment rather than empathy or connection.
- **Auditory Stimuli and Vibrations**: Witnesses also report hearing unfamiliar sounds during abductions, such as humming, beeping, or whirring noises. These sounds may accompany sensations of vibration or subtle pressure changes, adding another layer of sensory input. Some experiencers interpret these sounds as machinery or technology at work, reinforcing the idea that the abduction is part of a clinical process or experiment.

Physical Aftereffects and "Marks": Evidence of Contact?

A particularly compelling aspect of physical contact in abductions is the presence of physical marks or sensations that experiencers notice after the event. These can include unexplained bruises, scars, puncture marks, or sensations of soreness in specific areas of the body. For witnesses, these marks serve as tangible evidence that something beyond ordinary experience has taken place.

- **Unexplained Marks and Medical Anomalies**: Many experiencers report finding marks on their bodies following an abduction, often in patterns or shapes that seem unusual. These marks can include small punctures, triangular or circular bruises, or linear cuts that do not align with any known injury. Some experiencers interpret these marks as signs of medical testing or examination by the extraterrestrials, viewing them as physical remnants of their encounter.
- **Lingering Physical Sensations**: In addition to visible marks, some individuals describe residual sensations such as soreness, tenderness, or tingling in areas that were touched or examined during the encounter. These sensations often persist for hours or days, reinforcing the belief that a physical interaction took place. The presence of these sensations suggests that abductions involve both direct physical contact and sensory engagement, creating experiences that are hard to dismiss as mere hallucination or dream.

Memory Retrieval Methods: Hypnosis, Regression, and the Question of Validity

The intense sensory experiences reported in abductions are often fragmented, with witnesses struggling to remember the entire event. Many experiencers report having "missing time" or gaps in memory, which they later attempt to retrieve through hypnosis or regression techniques. These methods are commonly used to explore memories that feel obscured or repressed, allowing witnesses to access details of their experiences that they may have otherwise forgotten. However, the reliability of hypnosis and regression in memory retrieval is a subject of considerable debate, as these methods can sometimes introduce distortions or false memories.

The Role of Hypnosis in Uncovering Abduction Memories

Hypnosis is a popular tool for recovering lost or repressed memories, as it places individuals in a relaxed, focused state that can enhance recall. Many individuals who undergo hypnosis to retrieve abduction memories describe vivid, detailed accounts of their experiences, often recalling sensations and visuals that were previously inaccessible. This process can be both therapeutic and illuminating, providing witnesses with a coherent narrative of events that were previously fragmented or hidden.

- **Hypnotic Regression as a Memory Recovery Tool**: During hypnotic regression, an individual is guided to recall memories by focusing on sensations, emotions, or specific moments associated with the experience. In the context of abduction cases, regression can lead to the recovery of vivid sensory details, such as touch, sounds, and even emotions experienced during the encounter. This allows witnesses to construct a more complete memory, often providing clarity and insight into the experience.
- **Therapeutic Benefits and Emotional Release**: Hypnosis can also have therapeutic benefits, allowing witnesses to process any lingering trauma or fear associated with their abduction. By accessing hidden memories, witnesses may gain a sense of control over their experience, transforming the encounter from an overwhelming mystery into a comprehensible narrative. This therapeutic process can help experiencers cope with any lingering emotional impact, allowing them to reintegrate the memory in a way that feels meaningful.

The Validity of Hypnosis and Memory Distortion

While hypnosis and regression can unlock memories, these methods are not foolproof and are subject to criticism for their potential to introduce false memories. Memory is inherently malleable, and hypnosis can sometimes lead individuals to "remember" details that are influenced by suggestion or expectation rather than actual events. The reliability of these techniques is a topic of ongoing research, as psychologists seek to understand how hypnosis affects memory and whether it can reliably uncover factual details.

- **The Risk of Suggestion and Confabulation**: During hypnosis, individuals are more susceptible to suggestion, meaning that leading questions or subtle cues from the hypnotist can influence the details that are recalled. This phenomenon, known as confabulation, can create false memories or distort actual memories, making it challenging to distinguish between genuine recall and memories constructed under suggestion. In abduction cases, this can lead to embellishments or exaggerations that may not accurately reflect the experiencer's true memory.
- **Memory Fragmentation and Reconstruction**: Some researchers suggest that memories recovered through hypnosis may not represent a linear narrative but rather a reconstruction based on fragments of real and imagined details. Abduction encounters are often complex, involving intense sensory experiences that the brain may not fully process in real time. During hypnosis, the mind may attempt to "fill in the gaps," creating a coherent story that aligns with the experiencer's expectations or beliefs.

Understanding the Impact: Physical and Psychological Aftereffects of Abductions

The combination of physical sensations, marks, and memories recovered through hypnosis creates a lasting impact on abduction experiencers. Many report that the physical contact they experienced leaves them with lingering physical sensations or emotional scars, challenging their sense of reality and self. These encounters, whether real or imagined, have profound psychological effects, often leading individuals to seek answers and support in a world that may not believe or understand their experiences.

- **Long-Term Physical and Emotional Aftereffects**: Abduction experiences can create lasting physical and psychological aftereffects, with witnesses describing sensations, marks, or scars that persist long after the event. Emotionally, many experiencers struggle with feelings of isolation, confusion, or even fear, as they attempt to reconcile their memories with their understanding of reality. These aftereffects suggest that the sensory and physical aspects of abductions are deeply impactful, creating experiences that leave a permanent mark on both the body and mind.
- **The Search for Validation and Support**: Given the intensity of these experiences, many witnesses seek validation and support from others who have undergone similar encounters. Support groups, online forums, and even counseling services provide experiencers with a safe space to share their stories, offering comfort and solidarity in a community that understands. This quest for validation is often a central part of the healing process, as it allows witnesses to come to terms with their experiences and regain a sense of normalcy.

The Tangible Reality of Physical Contact and Memory in Abductions

Physical contact and abduction experiences introduce a profound and deeply personal dimension to extraterrestrial encounters. Through sensory experiences that engage the body, touch, and environmental stimuli, witnesses find themselves immersed in a reality that feels far beyond ordinary perception. These encounters create memories that are visceral and often unshakable, leaving physical marks, sensations, and emotional effects that linger long after the event has ended.

The process of memory retrieval, often achieved through hypnosis or regression, enables witnesses to access fragmented or hidden details of their experiences, creating a coherent narrative that can be both clarifying and transformative. However, the validity of these methods remains a topic of debate, as researchers question the accuracy of memories recovered through hypnotic techniques and explore the potential for suggestion or confabulation.

Ultimately, physical contact and communication through abductions challenge our understanding of memory, perception, and the limits of human experience. These encounters underscore the profound impact that sensory and physical elements have on shaping memory and meaning, highlighting the complexity of interactions that go beyond verbal language. Whether viewed as evidence of extraterrestrial contact or as a product of human consciousness, these experiences continue to captivate and mystify, inviting us to explore the boundaries of reality and the unknown.

Diverse Pathways of Connection in Extraterrestrial Encounters

The methods of communication reported in extraterrestrial encounters reveal a rich and multifaceted landscape of interaction that extends far beyond traditional human language. Through telepathic exchanges, symbolic imagery, and physical contact during abductions, each method of communication engages the experiencer on distinct cognitive, emotional, and sensory levels, offering a spectrum of messages that can resonate deeply and leave lasting impressions.

Telepathic communication stands out as an intimate, direct channel that bypasses linguistic boundaries, allowing experiencers to receive thoughts, emotions, and insights instantly. The role of telepathy in extraterrestrial messaging reveals the potential for mind-to-mind communication that feels both profound and unsettling, creating a sense of undeniable authenticity that often transforms the experiencer's worldview. However, the psychological impact of these experiences raises questions about perception, memory, and the power of suggestion, challenging researchers to explore how these telepathic messages shape and influence human consciousness.

Symbolic and visual communication presents another fascinating dimension, using images, symbols, and star maps to convey complex ideas. Unlike verbal or telepathic messages, these symbols offer a universal language that invites interpretation on multiple levels, blending intellectual, cultural, and spiritual themes. Yet, the interpretation of symbols introduces challenges, as experiencers bring their own cultural and psychological frameworks to the process, creating ambiguity and potential for subjective understanding. Through these visuals, extraterrestrials may be attempting to establish connection, orientation, or convey messages about cosmic order, which require us to look beyond literal meaning and engage with the symbolic depth of these encounters.

Physical contact and communication through abductions represent the most tangible and sensory aspect of extraterrestrial interaction. Abduction experiences often involve heightened sensations, touch, and physical aftereffects that ground these encounters in the body, making them feel deeply real. Witnesses frequently report physical marks, sensations of restraint, and environments that resemble medical examination rooms, leading to interpretations of scientific or experimental motives. Memory retrieval methods like hypnosis and regression help witnesses access and understand these encounters, but the reliability of these methods is subject to scrutiny, raising critical questions about the nature of memory, suggestion, and authenticity.

In exploring these diverse methods of communication, it becomes evident that extraterrestrial encounters engage the human mind and body in profound ways, challenging our understanding of reality, perception, and the boundaries of human experience. Each method—whether telepathic, symbolic, or physical—offers unique insights into how extraterrestrials might connect across the divide of species and consciousness, using pathways that provoke both wonder and reflection. Together, these methods form a complex and nuanced tapestry of interaction, inviting humanity to expand its perspective on communication, connection, and the mysteries that lie beyond the known world.

As we transition from examining the methods of communication in extraterrestrial encounters, it becomes essential to consider the broader psychological, sociological, and cultural contexts that shape these experiences. While the modes of interaction—telepathy, symbolism, and physical contact—offer insights into how messages are conveyed, the impact of these encounters extends far beyond the moment of communication itself. How individuals interpret, internalize, and share their experiences is profoundly influenced by psychological factors, social dynamics, and cultural backgrounds. Understanding the psychology of encounters, including trauma's effect on memory and cognitive models of perception, sheds light on how these experiences are processed internally. Simultaneously, mass hysteria and societal influences reveal how encounters can affect groups and communities, while cultural and religious frameworks shape the meaning assigned to extraterrestrial messages. By exploring these dimensions, we gain a richer perspective on the complex interplay between individual experience and collective understanding, providing a deeper appreciation for how extraterrestrial encounters resonate across both personal and cultural landscapes.

Chapter 4: Human Perspectives: Psychological, Sociological, and Cultural Dimensions of Extraterrestrial Encounters

Beyond the intriguing methods of communication in extraterrestrial encounters lies a compelling layer of human interpretation shaped by psychology, social dynamics, and cultural heritage. How people perceive, process, and understand these experiences varies widely, influenced by personal psychological factors, group interactions, and cultural beliefs. For some, encounters provoke profound, even traumatic memories that alter their understanding of reality, while others may interpret these events through the lens of societal or religious frameworks. The influence of mass hysteria and social conditioning also plays a role, as seen in documented group encounters that spark collective perceptions and responses. Cultural and religious backgrounds provide additional context, coloring interpretations with elements drawn from mythology, folklore, and spirituality. By delving into these perspectives, we uncover how deeply human factors influence the way encounters are internalized and shared, revealing an intricate web of meaning that transforms these experiences into phenomena that resonate across cultures and societies.

Exploring the Mind: The Psychology of Extraterrestrial Encounters

The psychological dimensions of extraterrestrial encounters are as complex as they are profound. Encounters often provoke strong emotional and cognitive responses that can linger for years, impacting how individuals perceive themselves and the world around them. While some view these experiences as enlightening or transformative, others find them deeply unsettling, even traumatic. Psychological factors such as trauma, memory processing, and cognitive interpretation play a significant role in shaping these experiences, providing a framework through which individuals attempt to make sense of what often feels beyond understanding.

In this exploration, we delve into the ways trauma impacts memory and perception in encounter experiences, examining the psychological effects that can shape or even distort memories. Additionally, we look at cognitive models that offer explanations for these extraordinary experiences, shedding light on how the human mind processes phenomena that defy conventional reality. By understanding these aspects, we gain insight into the powerful interplay between psychology and perception in encounters that challenge the limits of human comprehension.

The Impact of Trauma on Memory and Perception: Understanding the Emotional Fallout of Encounters

Encounters with extraterrestrials are often accompanied by intense emotions and sensations that can be traumatic, altering how individuals perceive and remember these events. Trauma affects memory in profound ways, sometimes causing fragments of memories to be repressed or distorted, while at other times, it leads to an exaggerated recollection of specific details. For individuals who experience trauma during encounters, this emotional response can create lasting psychological scars, making it difficult for them to fully process or understand what happened.

Trauma as an Amplifier: Heightened Emotions and Distorted Memories

Trauma has a powerful effect on the brain, amplifying emotions and altering the way memories are stored and recalled. In encounters with extraterrestrials, the overwhelming nature of the experience—whether due to fear, confusion, or awe—can heighten emotions to such an extent that the brain encodes memories in a fragmented or exaggerated way. These intense emotions often lead to vivid, intrusive memories that the experiencer may struggle to reconcile with their everyday life.

- **Fragmented and Vivid Recall**: Traumatic encounters can lead to fragmented memories, where certain aspects of the experience are recalled in vivid detail, while other parts are entirely blank. This fragmentation occurs because trauma impacts the brain's hippocampus, the area responsible for organizing and storing memories. The result is often a memory that feels incomplete, with certain sensations or images standing out while others are inaccessible. This fragmented recall can create confusion, leading experiencers to question their own perception and reality.
- **Hyperfocus on Specific Details**: In traumatic encounters, the brain sometimes fixates on specific details, such as the color of an extraterrestrial's eyes or the sound of machinery. These details become lodged in the experiencer's mind, replaying in a loop that can lead to obsessive thoughts or flashbacks. This hyperfocus may be the brain's way of attempting to make sense of a reality-defying experience, emphasizing details that seem important while pushing other parts of the memory into the background.

Psychological Aftereffects: Fear, Anxiety, and PTSD-like Symptoms

The trauma of an extraterrestrial encounter can leave lasting psychological effects, similar to post-traumatic stress disorder (PTSD). Many experiencers report symptoms such as heightened anxiety, difficulty sleeping, and recurring nightmares related to the encounter. These symptoms can have a profound impact on daily life, sometimes causing social withdrawal or even depression as individuals struggle to cope with an experience that feels beyond human understanding.

- **Nightmares and Flashbacks**: After a traumatic encounter, it is common for experiencers to have nightmares or flashbacks, where they relive parts of the event in unsettling detail. These recurring memories can be triggered by specific sights, sounds, or even smells that resemble aspects of the encounter. For some, these flashbacks are so intense that they feel as though they are re-experiencing the event, creating a sense of terror and helplessness.
- **Hypervigilance and Social Withdrawal**: Trauma often leads to hypervigilance, a state of heightened awareness and sensitivity to potential threats. Many experiencers of traumatic encounters find themselves constantly on edge, as though anticipating another encounter or bracing for the unknown. This state of hyperawareness can lead to social withdrawal, as individuals feel uncomfortable in crowds or around others who may not understand their experiences.

Cognitive Models for Encounter Experiences: Making Sense of the Unfathomable

Cognitive psychology offers various models to explain how the human mind processes encounters with extraterrestrials, particularly those that challenge our understanding of reality. Cognitive models examine how perception, memory, and mental frameworks interact to create an experience, especially when that experience lies outside the bounds of normal perception. By applying cognitive theories, we gain insight into how the mind may interpret, distort, or even fabricate elements of an encounter in response to factors like expectation, belief, and sensory overload.

Perception Under Stress: The Brain's Response to Anomalous Stimuli

When confronted with anomalous stimuli, the brain often struggles to make sense of the unfamiliar, filling in gaps or adjusting perceptions based on prior knowledge and expectations. This cognitive process is particularly relevant in extraterrestrial encounters, where unfamiliar shapes, sounds, and sensations can overwhelm the brain's capacity for interpretation, leading to unique mental phenomena such as pareidolia and perceptual distortion.

- **Pareidolia and Pattern Recognition**: The brain is naturally inclined to recognize patterns, a cognitive process known as pareidolia. In situations of heightened stress, such as an encounter, the brain may "see" patterns or faces where none exist, attempting to impose order on the chaotic or unknown. This phenomenon could explain why experiencers often describe extraterrestrials in human-like terms or perceive symbolic patterns in the environment, as the brain tries to categorize unfamiliar sights in a familiar framework.
- **Perceptual Distortion and Misinterpretation**: Under stress, the brain is prone to perceptual distortion, where sights, sounds, and other sensory information are perceived in exaggerated or altered ways. This distortion is often a result of adrenaline and heightened emotional states, causing individuals to misinterpret or magnify what they see and hear. In the context of an extraterrestrial encounter, this could mean that shapes are perceived as larger or more menacing, sounds are intensified, and time seems to stretch or condense.

Memory and Expectation: The Influence of Belief on Encounter Narratives

Expectations and prior beliefs can strongly influence how individuals experience and remember encounters. Cognitive theories suggest that memory is not a static record but a dynamic process influenced by beliefs, expectations, and cultural conditioning. When individuals expect to encounter extraterrestrials or have been exposed to alien-related media, their memories may be shaped by these expectations, altering how they interpret and recall the experience.

- **Expectation and the Power of Suggestion**: The mind's capacity to create or embellish memories based on expectation is a phenomenon well-documented in cognitive psychology. When individuals enter an experience with a strong expectation—such as anticipating contact with extraterrestrials—the brain may subconsciously fill in details that align with that expectation. This process can lead to memories that feel real but may incorporate elements influenced by prior knowledge or belief.
- **Cultural Conditioning and Memory Formation**: Culture plays a significant role in how experiences are perceived and remembered, particularly when it comes to extraordinary events. Individuals who have been exposed to media portrayals of extraterrestrial encounters may unconsciously draw on these representations when interpreting their own experiences. This cultural conditioning can lead to standardized "scripts" or recurring themes in encounter narratives, where details like the appearance of extraterrestrials or the sensation of being "abducted" follow familiar patterns.

Dissociation and the Mind's Defense Mechanisms

In extreme encounters that provoke intense fear or confusion, the brain sometimes employs defense mechanisms such as dissociation, where individuals feel detached from their own experiences. Dissociation can create a sense of unreality, where the experiencer feels as though they are observing the encounter from outside themselves. This psychological response often helps to protect the individual from overwhelming emotions, creating a mental barrier that separates them from the traumatic aspects of the event.

- **The "Unreal" Feeling: Dissociative Response in Encounters**: Dissociation can manifest as a feeling of being outside one's body, observing the encounter from a detached perspective. Many witnesses describe feeling "numb" or "detached" during encounters, as though the experience is happening to someone else. This sensation can make the memory feel dreamlike or surreal, adding to the experiencer's difficulty in distinguishing reality from imagination.
- **Long-Term Dissociative Effects**: For some, the dissociative response does not end with the encounter but continues as a coping mechanism. Individuals may feel disconnected from their emotions, have difficulty forming clear memories of the event, or struggle to integrate the experience into their sense of identity. Long-term dissociative effects can make it challenging to process or move on from the encounter, as the mind continues to shield itself from fully engaging with the memory.

Understanding the Psychological Impact of Extraterrestrial Encounters

The psychological landscape of extraterrestrial encounters is complex, involving trauma, memory distortion, cognitive biases, and defense mechanisms that all shape how individuals perceive and recall these experiences. Trauma often leaves fragmented or hyper-focused memories that feel vivid and real, while the brain's cognitive processes—such as pattern recognition, expectation, and dissociation—can influence the interpretation and recall of events. Together, these factors create a nuanced understanding of how the human mind navigates encounters that defy ordinary perception, revealing both the mind's adaptability and its limitations.

By examining the psychological impact of encounters through the lenses of trauma and cognitive psychology, we gain a deeper insight into the human response to extraordinary experiences. Whether encounters are viewed as authentic contact with extraterrestrial beings or as complex psychological events, understanding these processes highlights the profound influence of the mind on perception, memory, and meaning-making. Ultimately, this exploration underscores the intersection of psychology and the unknown, where the boundaries of reality and imagination blur in the face of experiences that challenge the limits of human understanding.

Collective Minds: Mass Hysteria and Social Influence in Extraterrestrial Encounters

Extraterrestrial encounters are not always solitary experiences. In some cases, they involve entire groups who report seeing the same phenomena, leading to a cascade of shared perceptions and beliefs. These group encounters raise complex questions about the role of mass hysteria and social influence, as well as the way media and societal factors shape how people interpret extraordinary events. When multiple people witness an encounter, the power of collective belief can amplify emotions and interpretations, creating a shared reality that transcends individual experience.

To understand these phenomena thoroughly, we examine notable cases like the Ariel School and Nuremberg events, exploring how group dynamics can create a cohesive narrative of an encounter. Additionally, we analyze the influence of media and societal narratives, considering how they shape our expectations and perceptions of extraterrestrial encounters. Together, these elements reveal the intricate interplay between individual perception, collective belief, and the cultural frameworks that influence how humanity processes the unknown.

The Dynamics of Group Encounters: Unraveling Mass Perception and Shared Reality

Group encounters with extraterrestrial phenomena offer unique insight into how people collectively perceive and interpret unusual events. These encounters, often marked by intense emotions and a strong sense of unity, show how mass perception can create a reality that feels undeniably true to those involved. In such situations, the lines between individual and collective experience blur, as group members reinforce each other's beliefs, shaping a narrative that resonates deeply within the group.

The Ariel School Encounter: Children, Fear, and Collective Belief

One of the most compelling examples of a group encounter is the 1994 incident at Ariel School in Ruwa, Zimbabwe, where over sixty children reported seeing a silver craft land near their school and a figure emerge from it. This figure reportedly communicated telepathic messages to the children about environmental destruction, warning them of future dangers. The story is notable not only for its content but for the sheer number of witnesses, all of whom were young children—a demographic generally regarded as credible due to their innocence and lack of preconceived notions about extraterrestrial encounters.

- **Fear and Fascination in a Shared Experience**: The children's shared fear and awe at the sight of the being and the craft created a powerful collective experience. Many described the event in similar terms, recounting details of the figure's appearance and the telepathic message they felt it conveyed. This sense of unity in their accounts underscores the psychological impact of shared emotion, as fear and fascination fueled each other, creating a bond among the children and reinforcing their belief in what they saw.

- **Collective Memory and Emotional Contagion**: The Ariel School incident highlights how emotional contagion—where emotions spread rapidly within a group—can influence memory and perception. The children's descriptions were strikingly similar, suggesting that, in the emotionally charged environment of the encounter, they may have unconsciously aligned their memories to create a cohesive story. This emotional contagion can amplify belief, as individuals take cues from one another to shape a shared narrative that strengthens over time.

The 1561 Nuremberg Celestial Event: A Historical Account of Collective Perception

Centuries before modern extraterrestrial narratives took root, an event in 1561 over Nuremberg, Germany, stunned the town's residents, who reported seeing a celestial battle involving strange shapes in the sky. Many described the spectacle as involving crosses, spheres, and cylinders, which they interpreted as a divine or supernatural sign. The event was so significant that it was chronicled in a woodcut and widely discussed, becoming part of the town's collective memory.

- **Religious Interpretation and Collective Belief**: In a deeply religious society, the people of Nuremberg interpreted the event as a warning or message from the divine, seeing it through a spiritual lens. This interpretation reflects how group beliefs are shaped by cultural and religious frameworks, which provide context and meaning to unfamiliar experiences. For the Nuremberg residents, the encounter was not

merely a strange occurrence; it became part of their shared understanding of morality and fate, reinforcing their belief in divine intervention.
- **Media of the Time: Chronicling the Event in Woodcut**: The woodcut of the Nuremberg Event served as the "media" of its time, preserving the memory of the event and sharing it beyond the town's borders. This early form of media coverage highlights how documentation can validate and perpetuate collective belief, reinforcing the narrative and allowing it to spread as a shared story. By engraving the event into history, the woodcut acted as a cultural anchor, ensuring that the collective memory of the encounter endured.

The Influence of Media and Society: Shaping Perception and Expectations

The role of media and societal narratives in shaping perceptions of extraterrestrial encounters cannot be overstated. In today's world, where mass communication reaches billions, the portrayal of extraterrestrials in films, television, and news has a profound influence on how people interpret strange phenomena. Cultural expectations, fueled by media portrayals, create a shared framework that primes individuals and groups to interpret experiences in specific ways, often mirroring popular extraterrestrial narratives.

Extraterrestrial Narratives in Media: Setting the Stage for Belief

From early science fiction novels to blockbuster films, media has played a crucial role in shaping the public's expectations of extraterrestrial encounters. Images of flying saucers, alien beings, and interstellar travel permeate popular culture, providing a visual and conceptual vocabulary for interpreting unexplained events. When individuals witness strange phenomena, they often draw on these media-influenced ideas to make sense of what they've seen, aligning their interpretations with culturally accepted narratives.

- **Media's Role in Shaping Extraterrestrial Archetypes**: The depiction of extraterrestrials in media has established common archetypes—such as the "little green men," the "greys," or benevolent beings from advanced civilizations—that shape public expectation. These archetypes create mental templates, influencing how people process and interpret encounters. Witnesses may unconsciously filter their perceptions through these archetypes, seeing what they expect to see or interpreting ambiguous elements in a way that aligns with popular representations.
- **The Feedback Loop of Media and Personal Belief**: Media does not just reflect public belief; it also reinforces it. When reports of extraterrestrial encounters are covered in the news or dramatized in film, they validate and perpetuate the narrative, creating a feedback loop. This loop reinforces collective belief in extraterrestrial life and conditions people to interpret unusual phenomena as evidence of extraterrestrial contact. In this way, media plays an active role in shaping and sustaining belief, fueling the cultural momentum of extraterrestrial narratives.

Social Media and the Spread of Collective Narratives

The rise of social media has amplified the impact of collective narratives, allowing encounter stories to spread rapidly and reach vast audiences. Platforms like YouTube, Twitter, and Reddit enable individuals to share their experiences instantly, often with video or photographic evidence that validates the story. Social media fosters an environment where collective narratives can grow quickly, as like-minded individuals connect, share, and reinforce each other's beliefs.

- **Viral Encounters and the Formation of Online Communities**: Social media provides a space for individuals with similar experiences to form communities, such as UFO enthusiasts or experiencer support groups. These communities create an echo chamber effect, where stories are repeated, validated, and embellished, contributing to a strong sense of belief among members. This environment encourages people to interpret encounters within a framework of collective validation, where each story reinforces the plausibility of the next.

- **Misinformation and the Role of Viral Content**: The rapid spread of information on social media can also lead to misinformation, as stories are shared without verification. This can lead to exaggerated or fabricated accounts that feed mass hysteria, as unverified claims quickly become part of the collective narrative. In this way, social media can distort perception, as stories that resonate emotionally or align with popular extraterrestrial tropes gain traction, influencing how individuals and groups interpret new encounters.

The Social and Psychological Mechanisms Behind Mass Hysteria

Mass hysteria, or collective psychogenic illness, occurs when individuals in a group experience similar physical or emotional symptoms without an identifiable cause. In the context of extraterrestrial encounters, mass hysteria can manifest as shared beliefs, fear, or even hallucinations, driven by the power of suggestion and social reinforcement. This phenomenon often arises in situations of high emotion or uncertainty, where people look to each other for cues on how to react, creating a self-reinforcing cycle of belief.

- **The Power of Suggestion in Group Settings**: In group encounters, suggestion plays a critical role in shaping perception, as individuals unconsciously align their beliefs with those of the group. When one person describes seeing an extraterrestrial or feeling a specific emotion, others in the group may adopt similar beliefs,

influenced by the power of suggestion. This creates a ripple effect, where each individual's belief reinforces the others, establishing a strong collective narrative.
- **Social Conformity and Reinforcement of Belief**: Social conformity also drives mass hysteria, as individuals in a group often adjust their perceptions to align with the majority. In encounters like the Ariel School incident, children may have felt pressure to conform to the descriptions given by their peers, shaping their memories to fit the group narrative. This need for conformity can make group encounters feel more cohesive, as individuals blend their experiences into a shared story that validates the beliefs of the whole.

Understanding Mass Hysteria and Social Influence in Extraterrestrial Encounters

Group encounters with extraterrestrials reveal the profound effects of social influence and mass hysteria on perception and belief. Through shared emotions and collective memory, groups can create powerful narratives that feel true to all members involved, as seen in cases like the Ariel School incident and the Nuremberg celestial event. These experiences demonstrate how fear, fascination, and cultural frameworks can amplify belief, creating a sense of shared reality that transcends individual perception.

Media and societal narratives further shape these beliefs, providing a foundation for interpreting encounters within culturally accepted extraterrestrial archetypes. The role of social media adds a new dimension, allowing encounter stories to spread rapidly and reinforcing collective belief through online communities. Together, these elements highlight the intricate ways that society and culture influence how we perceive and process encounters with the unknown, showing how easily individual experiences can transform into widely accepted collective realities. By understanding these dynamics, we gain a deeper appreciation for the psychological and social mechanisms that underlie mass encounters with the extraordinary.

Sacred Encounters: How Cultural and Religious Backgrounds Shape Interpretation of Extraterrestrial Messages

Extraterrestrial encounters and the messages they reportedly convey are rarely interpreted in a vacuum. Instead, these experiences are filtered through a tapestry of cultural beliefs, religious backgrounds, and personal values, leading to a wide range of interpretations that reflect the experiencer's worldview. While some view these encounters as purely scientific phenomena, others interpret them through spiritual or mystical lenses, seeing parallels with divine visions, religious epiphanies, or ancient myths. Understanding how cultural and religious influences shape the interpretation of extraterrestrial messages requires an exploration of human belief systems and the ways in which extraordinary events become interwoven with a culture's deepest values.

By analyzing how cultural backgrounds influence message interpretation and exploring the similarities between extraterrestrial encounters and mystical or religious experiences, we can gain a nuanced understanding of the diverse ways in which humans make sense of the unknown. This understanding not only sheds light on the psychological impact of such experiences but also reveals the universal human need to contextualize the extraordinary within familiar frameworks.

Cultural Backgrounds and the Filter of Belief: Shaping Interpretation of the Unfamiliar

Each culture provides its members with a set of beliefs, symbols, and stories that shape their understanding of the world. When individuals from different cultural backgrounds experience something as extraordinary as an extraterrestrial encounter, they are likely to interpret it through the lens of their own cultural narratives. These interpretations are often deeply personal, reflecting collective values, religious traditions, and mythological symbols that have been passed down through generations.

Interpreting the Alien as the Divine: Religious Overtones in Extraterrestrial Encounters

For many people, extraterrestrial encounters evoke powerful associations with religious experiences, leading them to interpret these messages as divine or supernatural. The extraterrestrial being is often perceived as a higher power, akin to angels or gods, delivering messages of warning, guidance, or prophecy. This tendency is particularly prevalent in cultures with strong spiritual traditions, where extraterrestrials are sometimes seen as messengers from the divine, reinforcing beliefs about the cosmos and humanity's place within it.

- **The Vision of an Otherworldly Messenger**: In cultures where spiritual visions and encounters with divine messengers are common, extraterrestrials may be viewed as celestial beings sent to impart wisdom or warnings. For example, in certain indigenous traditions, otherworldly beings are believed to come from the sky to communicate with shamans, and extraterrestrial encounters may be interpreted in a similar way. These beings are sometimes seen as symbols of transformation or protectors of the natural world, mirroring the spiritual guardians of traditional beliefs.
- **Moral Messages and Divine Warnings**: Many experiencers describe receiving messages from extraterrestrials that emphasize environmental stewardship, peace, or moral responsibility. For those with a religious background, these messages often resonate as divine warnings, reminiscent of religious

stories in which prophets are sent to deliver messages about humanity's moral failings. In this context, the extraterrestrials' messages are seen not as alien proclamations but as calls for repentance or spiritual awakening, similar to religious revelations.

Myth, Folklore, and Extraterrestrial Encounters: Continuities in Cultural Symbols

Cultural symbols and archetypes play a significant role in how people interpret the appearance and actions of extraterrestrials. Many cultures have ancient myths about beings from the sky, such as gods, spirits, or heroes who descend from the heavens. When an individual from one of these cultures encounters an extraterrestrial, they may draw on these traditional symbols, seeing the alien as an extension of their mythological heritage.

- **Gods, Heroes, and Sky Beings**: In numerous cultures, myths and legends describe beings who come down from the heavens, whether as benevolent gods or powerful heroes. For example, in the mythology of various Native American tribes, sky beings are seen as powerful figures who guide and protect humanity. When someone from such a culture encounters an extraterrestrial, they may interpret the being as a modern-day equivalent of these mythological sky beings, associating the encounter with themes of protection or revelation.
- **Extraterrestrials as Modern Archetypes**: Psychologically, extraterrestrials can be seen as modern archetypes that fulfill roles similar to ancient gods or heroes. They serve as symbols of the unknown, embodying human fears, hopes, and curiosity. Carl Jung, the renowned psychologist, suggested that UFOs could represent a projection of the collective unconscious, functioning as symbols that reflect humanity's quest for meaning and connection with something greater than itself. In this view,

extraterrestrials serve as the "gods" of the modern age, representing our desire for transcendence and understanding.

Parallels with Mystical and Religious Experiences: Finding Common Ground in the Extraordinary

Many elements of extraterrestrial encounters resemble mystical or religious experiences, from visions and voices to feelings of awe and transformation. This overlap has led some researchers to suggest that extraterrestrial encounters may be a modern expression of the same psychological and spiritual phenomena that underlie religious experiences. By examining the parallels between these encounters and mystical experiences, we can better understand how individuals interpret them through a spiritual lens and integrate them into their lives as meaningful, even sacred, events.

Visions, Voices, and Altered States: Common Elements in Extraterrestrial and Mystical Encounters

Experiencers of extraterrestrial encounters often report seeing visions, hearing voices, or entering altered states of consciousness that resemble mystical experiences. These sensory phenomena create a powerful emotional response, leading individuals to interpret their encounters as spiritually significant. In both extraterrestrial and mystical encounters, witnesses describe feelings of awe, humility, and wonder, often viewing the experience as a profound turning point in their lives.

- **The Visionary Experience**: Visions are a central component of many religious and mystical experiences, from the apparitions of saints to the visions of prophets. In extraterrestrial encounters, witnesses similarly report seeing beings or landscapes that appear otherworldly, often accompanied by intense colors, light, or shapes that seem to vibrate with energy. This visionary experience creates a sense of transcendence, leading experiencers to interpret the encounter as a glimpse into a higher realm or reality.

- **Hearing the Voice of the Otherworldly**: Another parallel is the experience of hearing a voice that communicates messages or guidance. In religious contexts, such voices are often attributed to angels, gods, or spirits, delivering messages to chosen individuals. In extraterrestrial encounters, witnesses sometimes hear telepathic messages or voices that convey warnings or instructions. This similarity between the "voice" of the extraterrestrial and the "voice" of the divine creates a bridge between secular and sacred interpretations, allowing witnesses to view the experience as a form of divine communication.

Transformation and Revelation: The Lasting Impact of Contact Experiences

Both extraterrestrial and mystical encounters often lead to personal transformation, with experiencers describing a shift in their worldview, values, or sense of self. This transformative aspect is a hallmark of religious conversions and mystical awakenings, where the individual feels forever changed by the encounter. For those who interpret their extraterrestrial experience as a form of revelation, the impact can be life-altering, leading to a newfound sense of purpose, humility, or spiritual insight.

- **Spiritual Awakening and Cosmic Awareness**: Many witnesses of extraterrestrial encounters describe feelings of cosmic awareness, seeing themselves as part of a larger, interconnected universe. This perspective is also common in mystical experiences, where individuals feel a profound sense of unity with the cosmos or a connection to a divine source. The extraterrestrial encounter thus becomes a form of spiritual awakening, inspiring individuals to re-evaluate their place in the universe and adopt new beliefs or practices that reflect their expanded awareness.

- **Existential Humility and the Desire for Meaning**: The experience of encountering an unknown being or realm often leads to existential humility—a realization of the vastness of the cosmos and the limits of human understanding. In religious and mystical contexts, this humility is often a pathway to devotion or a search for meaning. Similarly, individuals who have encountered extraterrestrials may feel a newfound sense of purpose or a desire to understand life's mysteries, leading them to explore spiritual paths or adopt ethical principles based on their encounter.

The Interplay of Culture, Religion, and Interpretation in Extraterrestrial Encounters

Cultural and religious influences profoundly shape how individuals interpret extraterrestrial encounters, offering them a familiar framework for making sense of the unknown. Whether viewed through the lens of traditional myths, divine messages, or mystical experiences, these encounters often take on meanings that resonate deeply within the experiencer's belief system. For those with strong religious or cultural backgrounds, extraterrestrials may appear as sky beings, divine messengers, or modern symbols of ancient archetypes, bridging the gap between past and present understandings of the otherworldly.

The parallels between extraterrestrial and mystical encounters reveal the universal human need for meaning and transformation. From visionary experiences to profound feelings of awe and humility, these encounters echo themes found in religious and spiritual narratives across cultures. By exploring the role of cultural and religious frameworks, we see how encounters with the unknown are integrated into the human story, allowing individuals to find purpose, connection, and insight in experiences that might otherwise defy comprehension. Ultimately, these interpretations highlight the deep-seated human desire to see the extraordinary not as isolated events, but as integral parts of a cosmic journey shared across time and culture.

The Interwoven Tapestry of Psychological, Sociological, and Cultural Perspectives in Extraterrestrial Encounters

The psychological, sociological, and cultural dimensions of extraterrestrial encounters reveal the profound complexity underlying how humans experience and interpret the unknown. Through the lens of psychology, we see how trauma, memory, and perception shape individual encounters, often creating fragmented memories that feel intensely real and impactful. Cognitive models offer insights into how people process these experiences, particularly when faced with phenomena that challenge their understanding of reality.

Group encounters and the phenomenon of mass hysteria highlight the powerful role of social influence. In cases like the Ariel School and Nuremberg events, we observe how shared emotions and societal influences create cohesive narratives that feel true to those involved, while media and societal frameworks further shape perceptions and reinforce collective beliefs. These encounters are not just personal experiences; they become shared stories, amplified by social reinforcement and shaped by cultural narratives that span generations.

Cultural and religious influences add another layer to these encounters, as individuals often interpret extraterrestrial messages through the lens of their spiritual or cultural backgrounds. Whether viewed as divine messengers, mythical sky beings, or symbols of a cosmic journey, extraterrestrials frequently embody the archetypes and themes that resonate within a culture's mythology or religious beliefs. Parallels with mystical and religious experiences underscore the universality of such encounters, illustrating humanity's deep-seated need to seek meaning, purpose, and connection within the extraordinary.

Together, these perspectives reveal the intricate web of factors that influence how we interpret encounters with the unknown. Extraterrestrial experiences are shaped not only by the mind and memory but also by the powerful currents of society, culture, and spirituality. By understanding these psychological, social, and cultural dynamics, we gain a richer appreciation of how humanity navigates the boundaries of reality, imagination, and belief in our collective journey to understand the cosmos.

As we move from the psychological, sociological, and cultural perspectives that shape individual and collective interpretations of extraterrestrial encounters, it is essential to explore how science approaches these mysterious experiences. Scientific investigation offers a critical framework for examining the validity, mechanics, and neurological aspects of message encounters, focusing on objective analysis to decode what might lie behind such phenomena. Hypnosis, for example, is often used to retrieve hidden memories from these experiences, though it remains a subject of intense debate regarding its accuracy and susceptibility to suggestion. Additionally, neurological studies shed light on how memory and perception can be influenced and distorted, particularly in high-stress or anomalous events, helping us understand how the brain processes encounters that may include telepathic or visual messages. Finally, research into telepathy and consciousness delves into parapsychology and emerging theories, such as quantum consciousness, in an attempt to find scientific footing for the often-reported phenomenon of mental communication in extraterrestrial encounters. Together, these scientific inquiries provide a vital counterbalance, allowing us to approach the unknown with both skepticism and curiosity, as we seek to bridge the gap between subjective experience and objective understanding.

Chapter 5: Decoding the Unknown: Scientific Approaches to Message Encounters with Extraterrestrials

Scientific investigation provides a rigorous lens through which we can examine the enigmatic nature of extraterrestrial message encounters. While individual experiences are often deeply personal and influenced by psychological and cultural factors, science strives to uncover objective insights into the mechanisms behind these phenomena. Hypnosis, for instance, is frequently employed to retrieve hidden memories from such encounters, but it brings with it a host of questions regarding reliability and the potential for memory distortion. Neurological studies further our understanding by exploring how the brain processes and sometimes distorts memories, particularly under unusual or high-stress conditions, offering clues about how perceptions of telepathic and visual messages might be shaped by our neurological framework. Additionally, research into telepathy and consciousness—ranging from parapsychology to quantum theories—pushes the boundaries of what we know about human cognition, delving into possibilities that might explain the telepathic connections reported in extraterrestrial encounters. This scientific exploration not only provides a critical grounding for understanding these extraordinary experiences but also opens up new pathways for considering how consciousness, memory, and perception intertwine in humanity's quest to make sense of the unknown.

Unlocking Hidden Memories: The Role of Hypnosis in Retrieving Messages from Extraterrestrial Encounters

Hypnosis has long fascinated scientists, psychologists, and the public alike for its ability to access parts of the mind that may be hidden from ordinary awareness. In the context of extraterrestrial encounters, hypnosis has become a widely used tool for retrieving memories of encounters that witnesses struggle to recall consciously. By guiding individuals into a state of deep relaxation and heightened focus, hypnosis aims to unlock details of these extraordinary events that may be otherwise obscured by trauma, cognitive suppression, or simple forgetfulness. However, the use of hypnosis to retrieve extraterrestrial messages is steeped in both intrigue and controversy, raising questions about the accuracy of these recalled memories, the potential for suggestion, and the ethical implications of relying on a technique that can influence memory construction. To gain a thorough understanding, we will explore the techniques of hypnosis, the reliability of hypnotically retrieved memories, and the controversies surrounding this powerful yet polarizing practice.

The Techniques of Hypnosis: Journeying into the Subconscious

Hypnosis is fundamentally about inducing a trance-like state in which individuals experience a relaxed and focused awareness that is distinct from their everyday consciousness. This state, often referred to as a "hypnotic trance," allows for heightened suggestibility and mental focus, which can enable individuals to recall memories or experiences that are not readily accessible in their waking state. Various techniques are employed to achieve this state, each designed to bypass the conscious mind's filters and reach deeper layers of memory and perception.

Induction Techniques: The Gateway to Hidden Memories

The initial phase of hypnosis, known as induction, is a critical step that prepares the subject to enter a relaxed and receptive state. Hypnotists use a range of techniques to guide individuals into a trance, with methods varying based on the hypnotist's training and the needs of the subject. Some of the most common induction techniques include:

- **Progressive Relaxation**: This technique involves guiding the subject to relax each part of their body sequentially, starting from the head or feet and working toward the opposite end. By focusing on relaxing individual muscles, the subject enters a deeper state of relaxation, which helps to quiet the conscious mind and facilitate access to the subconscious.
- **Eye Fixation**: This classic technique involves asking the subject to focus on a single point, such as a spot on the wall or the hypnotist's hand. As the subject's focus intensifies, their peripheral awareness begins to fade, leading them into a trance state where they become more receptive to suggestion.
- **Visualization and Guided Imagery**: Hypnotists may also use visualization techniques, asking the subject to imagine a calming environment, such as a beach or forest. This imagery helps deepen relaxation and create a sense of separation from the conscious mind, allowing memories to emerge more freely.

Deepening Techniques: Reaching Deeper Layers of the Mind

Once the subject has entered a light trance, hypnotists employ deepening techniques to guide them into a more profound state of hypnosis. In this deeper state, subjects are believed to be more capable of accessing repressed or forgotten memories. Common deepening techniques include:

- **Counting Down**: The hypnotist may ask the subject to imagine themselves descending a staircase or moving down a path, counting down with each step. This mental journey mimics a descent into the subconscious, symbolizing a move away from conscious awareness and into hidden memory.

- **Breathwork and Sensory Focus**: By instructing subjects to focus on their breathing or sensations within their body, hypnotists further enhance the subject's sense of inward focus. Breathwork helps to slow down mental activity, creating a quiet, receptive mental space where suppressed memories may emerge.

These techniques are carefully orchestrated to encourage memory retrieval without imposing external content, though the heightened state of suggestibility introduces an element of uncertainty about the purity of the recalled memories.

Reliability of Hypnotic Recall: Fact or Fiction?

While hypnosis is widely used to retrieve memories, its reliability is a matter of ongoing debate. Hypnotically retrieved memories can feel vivid and real, but they are not always accurate. The heightened suggestibility inherent in hypnosis can lead subjects to confabulate or blend real memories with imagined details, creating narratives that feel genuine but may not be fully grounded in factual events. Understanding the mechanisms behind hypnotic recall is crucial for assessing the reliability of memory retrieval in extraterrestrial encounters.

Memory Malleability: How Hypnosis Influences Recall

Human memory is not a static record of events but a dynamic process that is susceptible to distortion, suggestion, and alteration over time. When a person is in a hypnotic trance, their memory becomes even more malleable, which can result in both heightened recall and unintended fabrication.

- **Suggestibility and False Memory Formation**: Under hypnosis, individuals are highly suggestible, meaning that subtle cues from the hypnotist—whether intentional or not—can shape the memories they retrieve. In cases where a hypnotist asks leading questions, subjects may unknowingly create false memories to fulfill the implied expectation. This tendency is particularly concerning in extraterrestrial encounter cases, where the hypnotist's expectations or biases could influence the details the subject recalls.

- **Source Amnesia and Memory Blending**: During hypnosis, memories from various sources may blend together, creating a phenomenon known as source amnesia, where subjects lose track of the origin of specific details. As a result, subjects may recall scenes or elements from movies, books, or dreams, mistakenly attributing them to real experiences. This can make it challenging to determine which parts of a hypnotically retrieved memory are genuine and which may be products of imagination or suggestion.

Confabulation and the Power of Emotion in Hypnotic Recall

Hypnotically retrieved memories are often vivid and emotionally charged, which gives them a sense of authenticity. However, the emotional intensity of these memories does not necessarily indicate accuracy. Confabulation—the unintentional creation of false details to fill in memory gaps—often occurs under hypnosis, especially when the subject is motivated to "remember" the encounter as fully as possible.

- **Emotional Amplification and Perceived Validity**: When subjects recall memories under hypnosis, the heightened emotional state can amplify their belief in the memory's validity. This intensity can lead subjects to firmly believe in the reality of the experience, even if it contains confabulated elements. For experiencers of extraterrestrial encounters, this means that the recovered memories may feel deeply personal and vivid, creating a narrative that feels irrefutably real, even if some aspects are constructed.

Controversies Surrounding Hypnosis in Extraterrestrial Encounter Cases

The use of hypnosis in the context of extraterrestrial encounters is fraught with controversy. While many subjects report life-changing experiences that feel undeniably real, critics argue that hypnosis may introduce more fiction than fact. The primary concerns revolve around the potential for memory distortion, ethical issues regarding hypnotic suggestion, and the lack of consensus on how to verify hypnotically retrieved memories.

The Debate on Memory Distortion: How Much Can We Trust Hypnotic Recall?

One of the most contentious issues surrounding hypnosis is the possibility of memory distortion. While some practitioners argue that hypnosis can help unlock repressed memories, others contend that it may do more harm than good by leading subjects to believe in experiences that are partly or wholly constructed.

- **Research Findings on Hypnotic Memory Accuracy**: Studies on hypnotic recall have produced mixed results. While some research indicates that hypnosis can enhance memory accuracy, other studies suggest that it increases the likelihood of false memories. Research has shown that subjects under hypnosis may remember details with a high degree of confidence, but confidence is not a reliable indicator of accuracy. This discrepancy raises questions about how much weight we should give to hypnotically retrieved memories of extraterrestrial encounters.
- **The Influence of Expectations and Beliefs**: The hypnotist's expectations, as well as the subject's

beliefs, play a significant role in shaping the recall process. For instance, if a subject strongly believes in extraterrestrials, they may be more likely to interpret ambiguous memories as evidence of alien contact. Likewise, a hypnotist who believes in extraterrestrial encounters may inadvertently encourage the subject to recall details that align with popular extraterrestrial narratives. This interaction between belief and memory makes it challenging to establish objective truth.

Ethical and Practical Concerns: Is Hypnosis the Right Tool for Encounter Memory Retrieval?

The ethics of using hypnosis for memory retrieval are complex, particularly in cases involving extraterrestrial encounters. The heightened suggestibility and potential for memory distortion present ethical dilemmas regarding the responsibility of the hypnotist and the psychological impact on the subject.

- **The Risk of Psychological Harm**: Retrieving memories of extraterrestrial encounters can be psychologically taxing, especially if the subject recalls distressing or traumatic details. The potential for false memories means that subjects may experience intense emotions over events that may not have occurred. This risk raises ethical concerns, as subjects could suffer emotional distress based on memories that are not entirely factual.
- **Questions of Informed Consent**: Given the possibility of false memories and emotional distress, it is crucial for subjects to fully understand the risks involved in hypnotic memory retrieval. Informed consent should include a discussion of the limitations and potential pitfalls of hypnosis, allowing subjects to make an educated decision about whether to proceed with the process.

Navigating the Complexities of Hypnosis in Extraterrestrial Encounters

Hypnosis offers a unique pathway into the hidden layers of memory, but its use in retrieving memories of extraterrestrial encounters is complex and controversial. Through induction and deepening techniques, hypnosis seeks to unlock suppressed or inaccessible memories, creating vivid recollections that can feel profoundly real. However, the heightened suggestibility associated with hypnosis raises questions about the reliability of these memories. Memory malleability, confabulation, and emotional amplification can lead subjects to recall details that may not accurately reflect the original experience, particularly when suggestive questions or external expectations come into play.

The ethical concerns surrounding hypnosis add another layer of complexity, as the potential for psychological harm and memory distortion must be carefully weighed against the desire to access hidden memories. Ultimately, hypnosis serves as both a valuable tool and a contentious practice, offering a glimpse into the mysteries of human memory while reminding us of the delicate balance between exploration and influence. As we continue to investigate the role of hypnosis in extraterrestrial encounter cases, we are called to approach it with both openness and caution, recognizing that the search for truth may require us to question not only the memories we retrieve but the methods by which we retrieve them.

Uncovering the Past: Examining Documented Cases of Hypnosis in Extraterrestrial Message Retrieval

Hypnosis has played a central role in some of the most well-known cases of extraterrestrial encounters and message retrieval. For individuals struggling to recall details of their experiences, hypnosis provides a pathway to access memories that seem buried within the subconscious. Over the years, several high-profile cases have emerged in which hypnosis was used to uncover narratives of contact, abduction, and communication with otherworldly beings. These documented cases not only reveal the potential of hypnosis in exploring hidden memories but also highlight the complexities and controversies surrounding the method. As we delve into these documented accounts, we examine how hypnosis has been employed, what information was retrieved, and what these cases tell us about the potential and limitations of using hypnosis in extraterrestrial investigations.

The Case of Betty and Barney Hill: The First Major Abduction Narrative

The 1961 case of Betty and Barney Hill is perhaps the most famous early instance where hypnosis was used to uncover details of an alleged extraterrestrial encounter. Widely regarded as the first publicized alien abduction story, the Hills' case set the stage for countless subsequent narratives of extraterrestrial encounters. Their account, retrieved through hypnosis, became a model for abduction stories and remains one of the most thoroughly investigated cases to date.

The Encounter: A Journey into the Unknown

In September 1961, Betty and Barney Hill were driving through the White Mountains of New Hampshire when they claimed to have seen a bright light in the sky that followed their car. They reported feeling strange sensations and losing several hours of memory from that night, leaving them with a profound sense of unease. Betty experienced recurring nightmares about the event, which led the couple to seek the help of psychiatrist Dr. Benjamin Simon, a trained hypnotist.

Hypnosis Sessions and the Unfolding of Abduction Details

Over the course of multiple hypnosis sessions, Dr. Simon worked with the Hills to retrieve their memories of that night. Under hypnosis, Betty and Barney both recounted being stopped by beings who took them aboard a spacecraft. They described undergoing various physical examinations and receiving telepathic messages. Although Dr. Simon was initially skeptical, he noted that both individuals provided consistent details and seemed genuinely distressed by their memories, which suggested an authentic psychological experience.

- **Impact on the UFO Community**: The Hills' case became iconic in the field of ufology and established a common structure for abduction stories, including medical examinations, loss of time, and telepathic communication. The consistency of their story, retrieved through hypnosis, gave the account credibility in the eyes of many researchers, even as critics questioned the reliability of hypnotic recall.
- **Memory Fragmentation and Emotional Recall**: Betty and Barney's memories under hypnosis were marked by intense emotion and fragmented recollections, characteristic of trauma-induced memory recall. This fragmentation raised questions about the impact of trauma on memory retrieval, as well as the potential for confabulation under hypnosis.

The Travis Walton Case: The Forest Incident and Message from Beyond

The Travis Walton case of 1975 is another widely publicized abduction story that involved hypnosis for memory retrieval. Walton's experience differed from the Hills' in several ways, yet it reinforced the potential of hypnosis as a tool for uncovering hidden details in extraterrestrial encounters.

Walton's story became famous both for the unique circumstances of his alleged abduction and for the psychological implications of his retrieval process.

A Dramatic Disappearance and Public Speculation

Travis Walton was part of a logging crew working in the Apache-Sitgreaves National Forest in Arizona when he claimed to have been struck by a beam of light from a hovering craft. His coworkers witnessed the event, and Walton disappeared for five days, sparking a media frenzy and police investigation. Upon his return, Walton had no clear memory of what happened during his absence, prompting him to undergo hypnosis to retrieve his memories of those five days.

The Hypnosis Sessions: Recalling Encounters on a Spaceship

Under hypnosis, Walton recalled being taken aboard an extraterrestrial craft, where he encountered beings with large eyes who communicated telepathically. He also described an environment that was sterile and filled with strange machinery. Walton's account included interactions with both alien beings and human-like figures, leaving researchers puzzled about the nature of his experience.

- **Controversies and Public Skepticism**: While Walton's account received considerable media attention, critics questioned the credibility of his story, particularly given the highly publicized nature of his disappearance and return. Some suggested that Walton's memories under hypnosis could have been influenced by his experiences on Earth, either consciously or subconsciously.
- **Psychological and Emotional Themes**: Walton's account reflects common psychological themes in abduction stories, including feelings of vulnerability, disorientation, and existential reflection. These

emotional elements suggest that, regardless of the factual basis of the event, the memory retrieval process taps into deeply personal psychological material, making the recalled memories feel real and profound.

The Allagash Abduction: A Group Encounter and Shared Hypnotic Recall

In 1976, four men—Jim Weiner, Jack Weiner, Charlie Foltz, and Chuck Rak—claimed to have experienced an abduction while on a fishing trip on the Allagash Waterway in Maine. Known as the Allagash Abduction, this case is unique in that it involved multiple witnesses who all underwent hypnosis to retrieve suppressed memories of the event. The shared nature of their experience offers insight into how hypnosis can retrieve memories in group encounters and how these memories can reinforce each other's narratives.

Recovered Memories of a Collective Encounter

The men initially reported seeing a bright object in the sky and experiencing a period of lost time. Years later, each of them began having disturbing memories and dreams that hinted at an abduction experience. Hypnotherapist Dr. Raymond Fowler conducted hypnosis sessions with each of the men, during which they recalled being taken aboard a craft and subjected to various examinations.

- **Consistency in Group Recollections**: During hypnosis, all four men reported similar details, describing the appearance of the beings, the layout of the craft, and the nature of the examinations they underwent. The consistency in their memories, retrieved separately, added weight to their story and suggested that they had undergone a shared experience, whether real or imagined.
- **Potential for Shared Confabulation**: While the consistency of their memories was compelling, skeptics

argued that the men may have unconsciously influenced each other's recollections, particularly if they discussed the incident prior to hypnosis. This possibility highlights a key challenge in using hypnosis for group encounters: the difficulty in separating genuine shared memories from memories shaped by social or interpersonal factors.

Long-Term Psychological Effects on Participants

Following their hypnosis sessions, the men continued to report psychological effects, including vivid dreams, anxiety, and a sense of cosmic awareness. Their experiences under hypnosis had a lasting impact, shaping their beliefs about extraterrestrial life and their place in the universe. The Allagash Abduction illustrates how hypnotically retrieved memories can resonate deeply with individuals, influencing their worldview even when the factual basis of the memories remains uncertain.

Budd Hopkins and the Intruders Foundation: A Dedicated Approach to Hypnotic Retrieval

Artist and abduction researcher Budd Hopkins is known for his pioneering work with abductees, particularly through the Intruders Foundation, which he founded to investigate reports of alien abductions. Hopkins conducted hypnosis sessions with numerous abductees, documenting their experiences and analyzing the commonalities in their stories. His work was instrumental in establishing the abduction phenomenon as a recognized area of study within ufology.

Developing Patterns in Hypnotic Recall

Hopkins' work with abductees revealed several recurring themes in their hypnotically retrieved memories, including medical examinations, telepathic communication, and family histories of abduction experiences. His research suggested that abduction phenomena were not isolated incidents but part of a larger pattern that involved generations of families. Hopkins documented cases where hypnosis uncovered details that seemed consistent across multiple subjects, giving the phenomenon a broader and more systematic context.

- **Critiques of Hopkins' Methodology**: While Hopkins' research helped to legitimize abduction narratives, critics have argued that his enthusiasm for the phenomenon may have influenced the outcomes of his hypnosis sessions. Some have suggested that his framing of questions and expectations could lead to memory distortion or suggestibility, particularly given the vulnerability of subjects seeking validation for their experiences.
- **Building a Community of Abductees**: Through the Intruders Foundation, Hopkins provided abductees with a community of support, where they could share their experiences and explore their memories in a validating environment. This communal aspect reinforced the abductees' belief in the reality of their experiences and highlighted the powerful role that shared narratives play in shaping memory and interpretation.

Insights and Implications from Documented Hypnotic Retrieval Cases

The examination of documented cases of hypnosis in extraterrestrial message retrieval reveals both the power and limitations of this method. Cases like those of Betty and Barney Hill, Travis Walton, and the Allagash Four demonstrate how hypnosis can bring forth vivid and emotionally charged memories that feel intensely real to the experiencers. These cases underscore how hypnosis enables individuals to access seemingly hidden details that provide structure to ambiguous or incomplete memories of encounters.

However, the controversies surrounding these cases highlight the challenges of using hypnosis as a reliable tool for memory retrieval. Issues of suggestibility, memory distortion, and confabulation complicate the interpretation of hypnotically retrieved memories, especially in cases where external expectations or social dynamics may shape the recollections. Documented cases show that while hypnosis can create a powerful narrative framework for understanding unusual experiences, the memories retrieved under hypnosis may be influenced by psychological, interpersonal, and environmental factors.

Ultimately, these documented cases illustrate the delicate balance between exploration and skepticism that characterizes the use of hypnosis in extraterrestrial investigations. As we continue to investigate the role of hypnosis in message retrieval, it remains essential to approach each case with an understanding of both the potential insights it offers and the limitations that accompany hypnotic memory retrieval. The narratives of these individuals offer a profound glimpse into the mysteries of the human mind and the enduring quest to understand our place in a potentially populated universe.

The Fragile Fabric of Memory: Neurological Studies on Memory Distortion in Unusual Experiences

Memory, while crucial to our understanding of reality, is not an infallible record of events but a dynamic, reconstructive process vulnerable to distortion, especially under unusual or extraordinary circumstances. Scientific studies on memory distortion have shown that when individuals encounter strange or high-stress events, their recollection of these experiences can be particularly susceptible to errors, embellishments, or outright fabrications. For those who report encounters with extraterrestrial beings or messages, this susceptibility becomes especially relevant. Understanding how memory works in these scenarios can offer critical insights into why people recall extraterrestrial experiences so vividly, even when elements of these memories may not align with objective reality.

Through examining research on memory formation and the factors that contribute to memory distortion, we can gain a comprehensive understanding of how unusual experiences shape and, at times, distort memory. This exploration not only clarifies the neuroscience of memory but also highlights the implications for investigating extraordinary experiences—showing how memory, perception, and expectation intertwine to construct our understanding of what we believe we have encountered.

The Brain's Reconstruction Process: How Memory Forms and Fades

Memory is far from a static snapshot of events; rather, it is a reconstructive process that the brain continuously refines, reshapes, and sometimes alters each time we recall it. Neuroscientists describe memory as inherently malleable—a quality that allows us to adapt memories to changing contexts but also leaves them vulnerable to distortion. Understanding the way memory forms and changes over time is essential to exploring how people recall unusual experiences.

Encoding, Storage, and Retrieval: The Basic Building Blocks of Memory

The process of memory creation begins with encoding, where sensory information is transformed into a form the brain can store. This initial stage is influenced by attention, meaning that high-stress or highly unusual events are often encoded more vividly because they command our full focus. However, intense focus does not always equate to accuracy; under stress, certain details may be exaggerated or misinterpreted due to the brain's heightened state of arousal.

- **Emotional Amplification in Encoding**: Emotional experiences are generally encoded more deeply due to the involvement of the amygdala, a brain region that processes emotions. When individuals encounter something unexpected, like a potential extraterrestrial sighting, the amygdala intensifies the memory, making it more vivid but not necessarily more accurate. This amplification often contributes to the sense of realism and intensity that accompanies memories of unusual experiences.
- **Storage and the Role of Reconstruction**: Once encoded, memories are stored across neural networks throughout the brain. However, memories are not stored as discrete units but rather as interconnected fragments. Each time a memory is retrieved, it is reconstructed from these fragments, often incorporating new information, beliefs, or expectations into the recall process. This reconstructive quality makes memory both adaptable and vulnerable to distortion over time.

Retrieval: Memory Recollection as a Creative Process

When we recall a memory, the brain essentially reactivates the neural patterns associated with that event. This process can feel seamless, but in reality, each recollection is an opportunity for alteration, as memories are shaped by context, mood, and suggestion. In the case of unusual experiences, the memory retrieval process is particularly susceptible to influence from external factors, such as media depictions of similar events or discussions with others who have had similar experiences.

- **Constructive Recall and Confabulation**: During retrieval, the brain may fill in gaps in memory with details that make sense but are not necessarily accurate, a process known as confabulation. This is especially common in memories of extraordinary or highly emotional events, where missing details may be unconsciously filled in with culturally resonant elements—such as familiar tropes from UFO sightings or extraterrestrial encounters. These constructed elements can become integrated into the memory, leading individuals to believe that they genuinely experienced them.

The Vulnerability of Memory Under Stress: How Unusual Experiences Shape Recall

Stress and trauma are known to impact memory formation, particularly in terms of accuracy and detail. Unusual experiences, which often evoke stress or excitement, can thus create memory distortions that feel vivid yet may diverge from reality. Studies on memory under stress reveal that while people may remember the "core" of an event, the specific details are more likely to be altered or exaggerated.

The "Flashbulb Memory" Effect: Vivid but Not Always Accurate

Flashbulb memories are intense, detailed memories formed in response to emotionally charged events, such as witnessing something shocking or awe-inspiring. While these memories often feel crystal clear, research has shown that they are just as prone to error as other types of memories. The sensation of vividness may actually reinforce confidence in inaccurate details, creating a false sense of certainty.

- **Extraterrestrial Encounters as Flashbulb Memories**: People who report encounters with extraterrestrial beings or messages often describe their memories with a sense of clarity and vividness. These memories can feel like "flashbulb memories," etched into the mind in extraordinary detail. However, studies indicate that the subjective vividness of flashbulb memories does not guarantee accuracy, suggesting that even clear-seeming memories of extraterrestrial encounters may include distortions or embellishments.

Trauma, Dissociation, and Fragmented Recall

Highly unusual or frightening experiences can induce dissociative states, where individuals feel detached from reality. This detachment impacts memory encoding and can lead to fragmented recall, where only certain aspects of the event are remembered. Dissociation also heightens the likelihood of memory distortion, as the brain struggles to process and store the experience coherently.

- **Dissociation in Abduction Narratives**: Many people who report abduction experiences also describe feelings of dissociation, which may contribute to fragmented memories. For example, they might recall certain sensory details vividly, like the feeling of a cold metal surface, while other aspects of the encounter are hazy or missing. These gaps are often filled in through confabulation, leading to a memory that feels complete but is partially reconstructed.

Suggestion, Expectation, and the Power of Cultural Influence

Extraterrestrial encounters do not occur in a cultural vacuum; rather, they are influenced by societal narratives, media, and personal beliefs. Studies on memory distortion have shown that suggestibility and expectation play significant roles in shaping memories, particularly in the context of unusual experiences.

The Misinformation Effect: How External Information Alters Memory

The misinformation effect occurs when people's memories are altered by post-event information. For instance, if a person is told about details they supposedly "forgot," they may incorporate these elements into their memory, genuinely believing them to be true. This phenomenon is especially relevant in cases where individuals are exposed to cultural narratives or media depictions of extraterrestrial encounters, as these stories can shape the way they recall their own experiences.

- **Influence of Media on Memory of Encounters**: Cultural depictions of extraterrestrials often create expectations for how an encounter should look and feel. Studies have found that when individuals encounter unexplained events, they may unconsciously align their memories with culturally familiar patterns. In the context of extraterrestrial encounters, this means that even unique or ambiguous experiences might be remembered in ways that conform to familiar narratives, such as descriptions of "greys" or flying saucers.
- **Expectancy Effects and Memory Distortion**: Expectancy effects refer to the influence of prior expectations on memory recall. If an individual believes that extraterrestrials are benevolent or that abduction entails certain types of experiences, they may unconsciously shape their memory to align with these expectations. This effect underscores the importance of context in memory distortion, as culturally induced expectations can subtly guide memory reconstruction.

The Role of Neuroimaging in Memory Distortion Research

Advances in neuroimaging technology have allowed researchers to study memory processes at a neurological level, shedding light on how unusual experiences can lead to memory distortion. Techniques such as fMRI (functional magnetic resonance imaging) and PET (positron emission tomography) provide a window into the brain's activity during memory recall, revealing the underlying mechanisms of memory distortion.

fMRI Studies: Tracking Memory Construction in Real Time

Functional MRI studies have shown that the brain regions involved in memory recall are also active during imagination and visualization, suggesting that memory and imagination may not be as distinct as we assume. When recalling a memory, the brain "replays" patterns of neural activity associated with the original experience but may also activate areas associated with creativity and imagination, leading to memory embellishment.

- **The Overlap of Memory and Imagination**: Studies using fMRI have demonstrated that memory recall and imaginative processes activate similar areas in the brain, including the hippocampus and prefrontal cortex. This overlap can lead to the blending of real and imagined elements, especially in memories of unusual experiences. For individuals recalling extraterrestrial encounters, this neurological blending may create a memory that feels authentic but includes constructed or exaggerated details.

PET Imaging and Emotional Intensity in Memory Recall

PET imaging has revealed that emotionally charged memories, like those formed during unusual or stressful events, show heightened activity in the amygdala and hippocampus. This heightened activity reinforces the vividness of the memory, making it feel more concrete and memorable. However, the strong emotional component can also lead to memory consolidation based on subjective experience rather than objective fact, making the memory more susceptible to distortion over time.

- **Emotion-Driven Memory Consolidation**: PET studies have shown that emotionally intense memories are often consolidated more deeply, meaning they are retained for longer periods. However, this consolidation is based on the emotional significance rather than the accuracy of the memory's content. This finding suggests that individuals recalling extraterrestrial encounters may remember the emotions of the experience more accurately than the specific details, which could lead to memory distortion.

Memory Distortion and the Neuroscience of Unusual Experiences

Research on memory distortion in unusual experiences reveals the complex interplay of perception, emotion, and external influence in the process of recall. Memory, especially under extraordinary circumstances, is not a flawless record but a flexible, evolving narrative that can be shaped by emotional intensity, cultural influences, and neurobiological processes. For those who report encounters with extraterrestrial beings, this dynamic nature of memory underscores the potential for both vividness and distortion, as the mind reconstructs events that feel deeply real yet may incorporate imaginative or suggestive elements.

Neurological studies using fMRI and PET imaging reveal that memory recall involves networks associated with both memory and imagination, highlighting the brain's reconstructive nature. This blending of fact and fiction, amplified by emotional arousal and suggestibility, contributes to the vividness and conviction often seen in reports of extraterrestrial encounters. While these memories are compelling, understanding the factors that contribute to memory distortion offers a grounded perspective, reminding us that the experience of memory is powerful and authentic even when it may not be entirely accurate. In the quest to understand extraordinary experiences, insights from neuroscience allow us to appreciate both the depth of human perception and the limitations inherent in our recollections.

The Brain's Role in Decoding the Unseen: Understanding How We Process Telepathic and Visual Messages

The brain is our primary instrument for interpreting reality, and this becomes particularly intriguing when we consider claims of telepathic and visual messages from extraterrestrial encounters. Neuroscientists have long explored how the brain processes sensory information, and in recent years, research has delved into how our perception might accommodate nontraditional forms of communication, such as telepathic and visual phenomena. When individuals report receiving mental images or direct "thought transmissions" from extraterrestrial entities, they are describing experiences that challenge conventional understanding of perception and memory.

Studying how the brain interprets these messages offers insights into both the nature of perception and the complex ways in which the brain constructs reality. Through examining the neurological mechanisms that could underlie telepathic experiences and visual encounters, we can begin to understand how such phenomena might be experienced, remembered, and integrated into human consciousness. This exploration spans from understanding brain regions involved in processing external sensory data to speculative inquiries into consciousness and its potential extensions beyond ordinary perception.

Interpreting Telepathic Messages: How the Brain Processes Nonverbal Communication

The idea of telepathy, or "mind-to-mind" communication, is often relegated to science fiction, but many who report encounters with extraterrestrial beings describe receiving thoughts or messages directly into their minds. Neurologically, this raises fascinating questions: Could the brain process information that originates from outside traditional sensory channels? While direct telepathy lacks substantial empirical support, studies on similar brain processes—like empathy, intuition, and nonverbal communication—offer clues about how the brain might interpret a "message" that doesn't rely on spoken language.

The Mirror Neuron System: A Mechanism for Mind-Reading?

Mirror neurons, specialized cells in the brain, fire both when we perform an action and when we observe someone else performing that same action. Originally identified in primates, mirror neurons are believed to play a significant role in empathy and social cognition by enabling us to "mirror" the mental state of others. This system could provide a neurological foundation for understanding telepathic experiences, as it enables us to interpret others' emotions and intentions even without explicit communication.

- **Empathy as a Neural Foundation for Telepathy**: Mirror neurons help humans intuitively grasp others' emotions, suggesting that the brain is equipped to sense mental states nonverbally. For those reporting telepathic messages, it is possible that the mirror neuron system is highly activated, enabling them to "sense" the emotions or intentions of an entity in ways that feel akin to telepathy. While this doesn't confirm the existence of telepathy, it illustrates the brain's

capacity to perceive intentions nonverbally, potentially giving rise to experiences that feel like mind-to-mind communication.
- **The Role of the Temporal Parietal Junction**: This brain region is crucial for theory of mind, or the ability to understand that others have separate thoughts and emotions. Activations in this region could support an individual's perception of "receiving" a message, as it allows the brain to model the mental state of another, making it plausible for experiencers to interpret their own intuitions or emotions as coming from an external source.

Brainwave Synchronization and the Possibility of Shared Mental States

Brainwave synchronization, or neural entrainment, occurs when two individuals' brainwaves become aligned, often during activities like meditation or deep interpersonal connection. Studies have shown that when people are emotionally or mentally attuned to one another, their brainwaves may become more synchronized. In reported extraterrestrial encounters, telepathic communication could be explained by this phenomenon, where individuals feel they are "on the same wavelength" as another consciousness.

- **Theta Waves and Altered States of Perception**: Theta brainwaves, associated with deep relaxation and heightened intuition, are prominent in meditative or trance-like states. For experiencers of extraterrestrial encounters, entering a state of high theta activity might enhance their sensitivity to subtle signals or perceptions, contributing to a sense of receiving telepathic messages. Theta states are also associated with vivid imagery and intuitive thought, aligning with descriptions of telepathic communication.

Visual Messages: How the Brain Constructs Otherworldly Images

Visual messages—mental images or visual impressions seemingly transmitted from an external source—are another common aspect of extraterrestrial encounters. These images can range from abstract symbols to vivid scenes, and experiencers often interpret them as intentional communications. Neurologically, this raises questions about how the brain might produce such images without external visual stimuli and how it imbues them with meaning. To understand this, we must explore the brain's visual processing system and its role in generating mental imagery.

The Occipital Lobe and the "Seeing" Brain

The occipital lobe, located at the back of the brain, is responsible for processing visual information. Interestingly, the brain's visual cortex is not only active when we see things with our eyes but also when we imagine or remember visual scenes. This ability for "inner vision" could explain how individuals might experience vivid visual messages that feel real even though they arise internally.

- **Mental Imagery as a Constructed Experience**: Neurologically, mental imagery is a "constructed" experience—meaning the brain can generate images as vividly as those seen with the eyes, especially during dream states or moments of heightened imagination. For those encountering visual messages in extraterrestrial experiences, their brain might be constructing images based on internal signals or subconscious cues, allowing them to see scenes that feel like they are being "sent" from an external source.
- **Occipital and Parietal Interaction in Visual Messages**: The parietal lobe, responsible for spatial awareness, works with the occipital lobe to generate a cohesive visual experience. This interaction helps

explain how visual messages might include spatial details or symbols with significance, as the parietal lobe organizes these images in a coherent and meaningful way. When individuals perceive symbols or landscapes as part of a visual message, this parietal-occipital collaboration might be at work, shaping abstract impressions into meaningful images.

Activation of the Default Mode Network: The Brain's Daydreaming Center

The brain's default mode network (DMN) becomes active during periods of rest, imagination, and self-reflection. This network, which includes areas such as the medial prefrontal cortex and posterior cingulate cortex, is responsible for spontaneous thoughts and mental imagery. In the case of visual messages during extraterrestrial encounters, the DMN might facilitate the spontaneous generation of images, which the experiencer then interprets as an external communication.

- **The Role of the DMN in "Downloading" Visual Messages**: When experiencers describe receiving visual messages, they may be accessing the DMN's capacity for spontaneous image generation. In a relaxed or trance-like state, the DMN becomes active, allowing the brain to "download" images that arise from subconscious processes. The experiencer may interpret these images as having been externally transmitted, especially if they feel symbolically significant or emotionally charged.

The Influence of Expectations, Memory, and Culture on Interpreting Messages

While the brain plays a fundamental role in generating telepathic and visual messages, expectations, memories, and cultural influences significantly shape how individuals interpret these experiences. The brain does not simply passively receive information; it actively filters, contextualizes, and integrates perceptions based on prior knowledge and beliefs. This interpretive process affects how experiencers perceive telepathic and visual messages, especially in contexts that involve extraterrestrial or otherworldly elements.

Memory and Expectation: How the Brain Constructs Meaning

Memory is an active process, influenced by expectation and previous experience. For individuals who anticipate certain types of messages or who have encountered extraterrestrial imagery in popular culture, the brain may unconsciously shape their experience to align with these expectations. When individuals report receiving telepathic or visual messages, their memories of the encounter may integrate familiar symbols or ideas that resonate with cultural narratives about extraterrestrials.

- **Schemas and Cultural Archetypes**: Schemas—mental frameworks built from experience—guide perception and memory. For instance, people who believe in extraterrestrials may have a mental schema that includes telepathy or symbolic imagery as part of the encounter. When they perceive unusual mental or visual phenomena, the brain may interpret these as messages, constructing a narrative that fits within this schema.
- **The Role of Priming in Interpreting Visuals**: Priming, a psychological effect where exposure to certain stimuli influences perception, can shape how visual messages are interpreted. If an individual has been exposed to specific images or ideas related to extraterrestrials, they may be more likely to perceive ambiguous visual impressions as messages, as the brain primes them to recognize and assign meaning to such images.

Synesthesia and Cross-Modal Perception: Merging Sensory Experiences

Synesthesia, a phenomenon where one sensory experience evokes another (such as seeing colors in response to sounds), reveals the brain's capacity for cross-modal perception. In cases of visual or telepathic messages, synesthetic-like processes could explain how certain images or sensations are experienced as external messages. If an individual has a heightened capacity for cross-modal perception, they may be more likely to experience abstract thoughts or emotions as concrete images or voices, giving the impression of receiving a "message."

The Brain as Interpreter of Extraordinary Experiences

Understanding how the brain interprets telepathic and visual messages reveals a complex interplay between neural mechanisms, memory, expectation, and sensory processing. While there is no definitive scientific explanation for telepathic communication, research into empathy, brainwave synchronization, and mirror neurons offers potential foundations for understanding how such experiences might be perceived as real. Similarly, visual messages can be explained by the brain's capacity to generate mental imagery and spontaneous symbols through interactions among the occipital lobe, parietal lobe, and the default mode network.

Expectations and cultural influences further shape how individuals perceive and recall these experiences, as memories and interpretations are actively constructed to align with preexisting beliefs and societal narratives. The brain's remarkable adaptability and its ability to merge sensory experiences suggest that extraordinary encounters—whether telepathic, visual, or otherwise—are rooted in a fascinating combination of perception, interpretation, and imagination. In exploring the neurological foundations of these experiences, we gain deeper insight into the brain's role as both a receiver and creator of the messages that individuals interpret as coming from beyond the ordinary boundaries of human communication.

Telepathy and the Mystery of the Mind: Exploring Parapsychological Research on Thought Transmission

The study of telepathy—often defined as the transfer of information from one mind to another without the use of the five traditional senses—has captivated researchers, mystics, and scientists for centuries. Within parapsychology, a field dedicated to investigating phenomena beyond conventional scientific explanation, telepathy occupies a central and highly contested space. Although telepathy lacks conclusive empirical support in mainstream science, numerous experiments, theories, and anecdotal accounts continue to fuel interest in the possibility that minds can communicate directly across physical space.

Exploring the parapsychological study of telepathy reveals a fascinating journey through groundbreaking experiments, theoretical developments, and philosophical debates. This examination also highlights the methodological challenges of studying telepathy, the controversies surrounding its legitimacy, and the broader implications of telepathic research for our understanding of consciousness itself. Parapsychologists have pioneered research methodologies aimed at capturing telepathic phenomena, pushing the boundaries of what is traditionally considered possible in human cognition.

The Foundations of Parapsychology: Telepathy as a Pillar of Inquiry

Parapsychology, formalized as a field in the late 19th and early 20th centuries, was established with the intention of scientifically investigating experiences that fall outside conventional understanding. Telepathy, alongside other "psi" phenomena such as clairvoyance, psychokinesis, and precognition, became a primary focus for early researchers, many of whom hoped to find empirical evidence that human minds could communicate without physical interaction.

The Early Days of Telepathy Research: The Society for Psychical Research

In 1882, the Society for Psychical Research (SPR) was founded in London to investigate paranormal phenomena using scientific methods. The SPR conducted some of the first formal studies on telepathy, developing protocols that aimed to separate genuine telepathic transmission from coincidence, suggestion, or fraud.

- **Initial Experiments and Anecdotal Evidence**: Early telepathy studies relied heavily on anecdotal evidence, including accounts of individuals who reported knowing the thoughts or emotions of loved ones at great distances. These accounts served as a starting point for more controlled experiments, as researchers sought to establish testable methods for studying telepathic communication.
- **Key Figures and Experiments**: Researchers like F.W.H. Myers and William Crookes conducted early studies on telepathy, often employing "sender" and "receiver" protocols in which one person attempted to transmit an image or thought to another in a separate location. Although these studies were preliminary, they laid the groundwork for the experimental frameworks that would later define telepathy research.

The Ganzfeld Experiments: A Controlled Attempt to Capture Telepathy

One of the most notable telepathy studies in parapsychology is the Ganzfeld experiment, which gained prominence in the

1970s and remains a widely cited approach. The Ganzfeld ("total field") method involves placing a "receiver" in a state of sensory deprivation, while a "sender" attempts to transmit images or thoughts. By minimizing sensory input, the experiment aims to create a mental state conducive to telepathic transmission.

- **Experimental Setup**: The receiver is seated in a dimly lit room, wearing headphones that play white noise and halved ping-pong balls over their eyes to create a homogeneous visual field. The sender, located in a separate room, is shown a randomly selected image or film clip and instructed to mentally "send" it to the receiver. The receiver then describes any images, thoughts, or impressions that come to mind.
- **Results and Controversies**: The Ganzfeld experiments yielded mixed results, with some studies reporting statistically significant evidence of telepathic communication, while others found no effect. Supporters argue that the positive results indicate a genuine telepathic phenomenon, while critics attribute the findings to methodological flaws, such as sensory leakage or statistical anomalies. Despite the controversies, the Ganzfeld experiments continue to be one of the most rigorously tested methods in telepathy research.

Theoretical Approaches to Telepathy: Is the Mind Boundless?

While empirical evidence for telepathy remains inconclusive, several theoretical frameworks have been proposed to explain how such a phenomenon could occur. These theories, which draw from quantum physics, consciousness studies, and psychology, provide a speculative foundation for understanding telepathy's mechanisms, suggesting that our current understanding of the mind and reality may be incomplete.

Quantum Entanglement and the "Nonlocal" Mind

Some parapsychologists argue that telepathy could be explained by principles of quantum physics, specifically the concept of "entanglement." In quantum mechanics, entangled particles remain connected in such a way that the state of one particle instantly affects the state of another, regardless of distance. Although quantum entanglement is typically observed at the subatomic level, researchers speculate that consciousness might operate similarly, allowing minds to connect beyond physical space.

- **The Hypothesis of Quantum Consciousness**: Theories of quantum consciousness, notably advanced by physicists like Roger Penrose and Stuart Hameroff, suggest that consciousness itself might be a quantum phenomenon. If the mind operates at a quantum level, it could potentially interact with other minds through entanglement, bypassing the limitations of physical separation. While this idea remains speculative, it provides a theoretical basis for telepathic communication.
- **Challenges and Criticisms**: Mainstream scientists often reject quantum explanations for telepathy, arguing that entanglement does not scale up to the level of human consciousness. Critics contend that using quantum theory to explain telepathy conflates physical processes with cognitive ones, and there is currently no empirical evidence to support this connection.

Collective Consciousness and Morphic Resonance

Another theory, proposed by biologist Rupert Sheldrake, posits that a "morphic field" or collective consciousness exists, through which information and experiences are shared across individuals. This concept, known as "morphic resonance," suggests that patterns of information—such as thoughts or emotions—can be transmitted telepathically across this shared field.

- **Morphic Fields as a Medium for Telepathy**: Sheldrake's theory implies that minds are not isolated but are part of a larger field that allows information to flow between individuals. In this view, telepathy is a natural outcome of interconnected consciousness, with thoughts and feelings "resonating" across a shared mental landscape.
- **Scientific and Philosophical Critiques**: Sheldrake's theory has been met with skepticism, as it lacks empirical support and conflicts with established biological and psychological models. However, it has found an audience among those who view consciousness as interconnected and holistic, resonating with ancient ideas about collective mind or universal consciousness.

Methodological Challenges and the Elusive Nature of Telepathy

The study of telepathy in parapsychology faces numerous methodological challenges, primarily due to the difficulty of capturing and measuring a phenomenon that lacks a clear physical basis. Despite these challenges, parapsychologists have developed various approaches to minimize bias, increase rigor, and test telepathy under controlled conditions. However, replicability and statistical rigor remain contentious issues.

The Problem of Replicability in Telepathy Experiments

One of the most significant challenges in telepathy research is the difficulty of replicating positive results. While some studies report statistically significant evidence for telepathy, others fail to reproduce these findings, leading critics to question the validity of the phenomenon.

- **Statistical Variability and the File Drawer Effect**: The "file drawer effect" refers to the tendency for studies with null results to remain unpublished, which can lead to an overrepresentation of positive findings in the literature. In telepathy research, the file drawer effect creates a biased view of the evidence, making it difficult to assess whether positive findings reflect genuine effects or statistical anomalies.
- **Efforts to Improve Replicability**: To address replicability concerns, parapsychologists have implemented rigorous protocols, such as preregistration of experiments, meta-analyses, and standardized testing environments. While these measures improve methodological rigor, the variability of results continues to be a barrier to mainstream acceptance of telepathy.

Controlling for Sensory Leakage and Expectation Bias

Sensory leakage—unintended transmission of information through sensory cues—poses a major challenge in telepathy experiments, as it can lead to false positives. Additionally, expectation bias, in which participants unconsciously align their responses with the experimenters' expectations, can further distort results.

- **Blind and Double-Blind Procedures**: To minimize sensory leakage and bias, many telepathy studies use blind or double-blind procedures, in which neither the participants nor the experimenters know the intended "message." These precautions reduce the likelihood of information contamination, though subtle biases may still influence results.

Philosophical Implications of Telepathy: Rethinking the Boundaries of Consciousness

Beyond the empirical challenges, telepathy raises profound philosophical questions about the nature of consciousness, individuality, and the mind's potential to transcend physical boundaries. Parapsychologists who study telepathy often grapple with these questions, considering the possibility that consciousness is not limited by the brain but may extend into a shared, nonlocal field.

The Mind as a Field of Consciousness: A Paradigm Shift

If telepathy exists, it would suggest that consciousness operates beyond the confines of individual minds, hinting at a "field" model in which all minds are interconnected. This idea challenges the traditional view of the brain as the sole generator of consciousness, suggesting instead that the brain is a receiver or transmitter within a larger field.

- **Dualism and Nonlocality**: Telepathy research aligns with dualistic theories that distinguish between the brain and consciousness, suggesting that mental phenomena are not entirely bound to physical processes. This perspective resonates with spiritual and philosophical traditions that view consciousness as a fundamental aspect of reality, existing independently of material boundaries.
- **Ethical and Existential Questions**: If telepathy is real, it raises ethical questions about privacy, autonomy, and the nature of human connection. The possibility of thought transmission challenges conventional ideas about individuality, prompting us to consider whether our minds are more interconnected than we realize.

The Quest to Understand Telepathy and the Limits of Human Perception

The exploration of telepathy in parapsychology offers a captivating glimpse into the mysteries of consciousness and the potential for direct mind-to-mind communication. While empirical support for telepathy remains limited and controversial, studies such as the Ganzfeld experiments and theories of quantum entanglement, morphic fields, and collective consciousness provide intriguing frameworks for understanding how telepathy might occur.

Methodological challenges—such as replicability, sensory leakage, and expectation bias—have hindered the mainstream acceptance of telepathy research. However, parapsychologists continue to refine their methods, striving for rigor and consistency in an attempt to capture what may be an elusive, yet deeply significant, phenomenon.

Telepathy's philosophical implications invite us to rethink our understanding of consciousness and the boundaries between individual minds. If telepathy were proven, it would suggest that our minds are not isolated but part of a larger, interconnected field of awareness, challenging us to expand our view of reality and reconsider the very nature of human connection. Whether or not telepathy exists, the study of this phenomenon enriches our understanding of consciousness, perception, and the vast potential of the human mind.

Quantum Consciousness: Bridging the Divide Between Mind and Matter

The theory of quantum consciousness represents a radical shift in how we understand the mind, suggesting that consciousness might operate at a quantum level rather than being a purely biological phenomenon confined to the brain. This concept has captivated both scientists and philosophers, as it could potentially explain phenomena like telepathy, consciousness beyond the brain, and even direct mind-to-mind communication. If consciousness operates on principles similar to those found in quantum mechanics, then our minds may not be as isolated as traditionally believed; instead, they could be interconnected through a web of nonlocal connections that transcend physical limitations.

To explore the possible links between telepathy and quantum consciousness theories, we will examine the foundational principles of quantum mechanics, leading theories on quantum consciousness, and the ways these theories intersect with parapsychological research. By delving into concepts like nonlocality, entanglement, and the observer effect, we can explore how quantum consciousness might provide a plausible framework for understanding telepathic communication.

Quantum Mechanics: A Brief Overview of Mind-Bending Principles

Before examining the specifics of quantum consciousness, it is helpful to understand the basic principles of quantum mechanics. Quantum theory, which describes the behavior of particles at the atomic and subatomic levels, has fundamentally challenged classical physics. Quantum mechanics is defined by principles that are often counterintuitive and seemingly contradictory to our everyday experiences, making it an ideal but controversial foundation for theories on consciousness and telepathy.

Wave-Particle Duality: The Fluid Nature of Reality

One of the key features of quantum mechanics is wave-particle duality, which holds that particles like electrons and photons can exhibit both particle-like and wave-like properties. This duality implies that subatomic particles do not have a fixed location or state until they are observed, existing instead in a probabilistic "superposition" of all possible states.

- **Implications for Consciousness**: If consciousness operates on similar principles, then it might exist in a state of "superposition," transcending fixed mental states and potentially allowing for multiple experiences or thoughts simultaneously. This could suggest that consciousness is not confined to the physical boundaries of the brain but is fluid and adaptable.

Quantum Entanglement: The Nonlocal Connection

Quantum entanglement occurs when two particles become so intertwined that the state of one particle instantly influences the state of the other, regardless of the distance between them. This phenomenon, famously called "spooky action at a distance" by Einstein, suggests that there is a level of reality in which distance is irrelevant.

- **Possible Role in Telepathy**: If human consciousness shares this entangled nature, then telepathic connections might arise from nonlocal entanglements between minds. This would mean that thoughts, emotions, or experiences could transfer instantly between individuals without the need for any physical medium. Entanglement in consciousness could therefore be the underlying mechanism that allows minds to communicate across distances, bypassing conventional sensory pathways.

Theories of Quantum Consciousness: Bridging Mind and Matter

Several theories have been proposed to explain how consciousness might operate at the quantum level. These theories are highly speculative and remain on the fringes of both neuroscience and quantum physics, yet they offer intriguing possibilities for understanding phenomena like telepathy.

Penrose-Hameroff's Orchestrated Objective Reduction (Orch-OR) Theory

One of the most well-known theories linking quantum mechanics to consciousness is the Orch-OR theory, developed by physicist Sir Roger Penrose and anesthesiologist Stuart Hameroff. The theory posits that consciousness arises from quantum processes within microtubules—tiny structures inside the brain's neurons. According to Orch-OR, these microtubules serve as quantum processors, enabling consciousness through quantum coherence and superposition states.

- **Microtubules as Quantum Channels**: Penrose and Hameroff argue that microtubules are structured in a way that allows them to support quantum states, which could produce the unique qualities of human consciousness. These quantum states may create a bridge between physical brain processes and the nonlocal properties of quantum mechanics, allowing for experiences that transcend time and space.
- **Telepathy Through Quantum Processes**: If consciousness indeed arises from quantum interactions within microtubules, it is conceivable that telepathic connections could be facilitated by these quantum states. Under this model, two individuals' consciousnesses could become "entangled" at a quantum level, allowing for direct thought transmission. This theory provides a theoretical framework for understanding telepathy as a product of quantum coherence in the brain, although empirical evidence for Orch-OR remains limited.

The Quantum Brain Hypothesis: A Nonlocal Network

Another approach to quantum consciousness is the Quantum Brain Hypothesis, which posits that the brain operates as a quantum system, using quantum computation to process information. This theory suggests that consciousness may be a field or wave function that interacts with the brain rather than being produced by it, giving consciousness a "nonlocal" nature.

- **Consciousness as a Field**: According to the Quantum Brain Hypothesis, consciousness might exist as a field that is not bound to the brain but can extend beyond it, interacting with other fields or waves of consciousness. This would mean that consciousness is not confined to individual minds but is part of a larger, interconnected field. Telepathic communication could then be understood as a resonance or interaction within this shared field.
- **Nonlocality and Mind-to-Mind Communication**: Nonlocality in quantum mechanics implies that distance does not prevent communication between entangled entities. In the context of consciousness, this could mean that minds are capable of instantaneous connections, even across vast distances. Telepathy could arise as a natural function of consciousness' nonlocal properties, as minds interact within a shared quantum field.

The Observer Effect: Consciousness and the Act of Observation

The observer effect in quantum mechanics refers to the fact that the act of observing a quantum system appears to affect its state. This principle has fueled speculation that consciousness itself might play a role in shaping reality, as it implies that reality is, in some sense, "co-created" by conscious observation. This concept has profound implications for our understanding of the mind and its potential role in telepathy.

Reality as a Product of Conscious Observation

The observer effect suggests that conscious beings do not merely perceive reality—they actively shape it through observation. This has led some researchers to propose that consciousness might interact with reality on a quantum level, influencing the physical world in ways that go beyond traditional sensory experience.

- **Telepathic Communication as a Co-Created Reality**: If consciousness indeed shapes reality, telepathy might be understood as a shared mental creation, in which two individuals co-create a mental connection that enables direct communication. This shared reality would not require physical transmission; instead, it would emerge from the mutual act of "observing" a telepathic connection, allowing thoughts and emotions to be shared as if by agreement within a shared consciousness.
- **Quantum Consciousness as a Gateway to Telepathic Experiences**: The observer effect implies that reality is malleable and influenced by consciousness. If telepathic experiences occur as part of this consciousness-driven reality, it may suggest that such experiences are more than mere anomalies—they could be an intrinsic part of how consciousness interacts with itself, potentially enabling minds to connect and communicate across perceived barriers.

The Challenges of Studying Quantum Consciousness and Telepathy

Quantum consciousness remains a speculative theory, largely because consciousness does not easily fit within the empirical frameworks of physics or neuroscience. Telepathy, in particular, is difficult to test through traditional scientific methods due to its elusive, subjective nature. However, several challenges define the ongoing exploration of quantum consciousness and telepathy.

The Problem of Measurement: Quantum and Subjective Phenomena

One of the most significant obstacles in studying quantum consciousness and telepathy is the problem of measurement. Quantum states are notoriously difficult to observe, as they exist in probabilistic superpositions until measured. Consciousness, similarly, resists objective measurement, as it is inherently subjective and often bound to personal experience.

- **The Paradox of Objectivity in Consciousness Studies**: Quantum consciousness research highlights the difficulty of applying objective scientific methods to subjective phenomena. The study of telepathy compounds this issue, as direct mind-to-mind communication is difficult to quantify. This measurement problem has led researchers to explore novel methodologies, including phenomenological studies and indirect measurements, to capture evidence of telepathy.

The Need for Interdisciplinary Approaches

Understanding quantum consciousness and telepathy likely requires an interdisciplinary approach, bridging physics, neuroscience, psychology, and philosophy. Theories on quantum consciousness challenge traditional boundaries between fields, requiring input from multiple disciplines to fully explore the implications of a nonlocal, quantum-based mind.

- **Collaborative Research as a Path Forward**: The complexity of quantum consciousness suggests that no single field can fully explain it. By combining insights from physics, psychology, and cognitive science, researchers hope to gain a more comprehensive understanding of telepathy and consciousness. This interdisciplinary approach may also provide new ways to test and validate theories on the quantum nature of the mind.

Quantum Consciousness and the Potential Reality of Telepathy

Quantum consciousness theories offer a tantalizing glimpse into the possibility that the mind operates on principles similar to those governing subatomic particles. Through concepts like nonlocality, entanglement, and the observer effect, these theories challenge conventional views on the mind's boundaries, suggesting that consciousness could transcend individual experience and interact across space in ways we have yet to fully understand.

Theories such as Penrose-Hameroff's Orch-OR and the Quantum Brain Hypothesis provide speculative yet fascinating frameworks for understanding how telepathy might occur. By proposing that consciousness operates within a quantum field or wave function, these theories open the door to the idea that telepathic connections might arise naturally within an interconnected consciousness. However, the inherent difficulties in measuring and studying quantum consciousness mean that these theories remain largely theoretical, grounded more in philosophical speculation than empirical proof.

As we continue to explore quantum consciousness, the potential links between telepathy and quantum mechanics invite us to reconsider the nature of reality and the mind's role within it. If consciousness is indeed nonlocal and capable of transcending physical limitations, telepathy could be an expression of a fundamental interconnectedness that science is only beginning to grasp. Whether or not these theories ultimately prove telepathy's reality, the study of quantum consciousness enriches our understanding of the mind, offering new possibilities for what human experience—and connection—might entail.

Summary of Scientific Investigations into Message Encounters: Hypnosis, Memory, and the Boundaries of Consciousness

The scientific exploration of message encounters from extraterrestrial beings has led researchers across disciplines to examine the mechanisms underlying human perception, memory, and consciousness. By investigating these phenomena through methods like hypnosis, neurological studies, and theories of telepathy, scientists and parapsychologists aim to understand how individuals process, recall, and interpret extraordinary experiences.

Hypnosis has emerged as a controversial yet influential tool in retrieving details from unusual encounters, such as abduction experiences. Techniques like regression hypnosis offer a means to access deeply embedded memories, though questions about reliability and the potential for suggestibility complicate the field. Documented cases show that hypnosis can lead to vivid recollections; however, these memories are prone to distortion, shaped by both the psychological state of the individual and the context of the experience itself.

Memory research further highlights the brain's susceptibility to error, particularly under conditions of stress or trauma. Studies reveal that memories are not static records but fluid constructs, influenced by emotions, expectations, and cultural narratives. This understanding is especially relevant to extraterrestrial encounters, where memory distortions may contribute to vivid but altered recollections of telepathic messages or visual experiences. Additionally, the brain's role in processing unconventional messages—be they telepathic or symbolic—reveals a complex interplay between perception and cognition, where the line between internal and external realities blurs.

Finally, studies on telepathy and quantum consciousness expand the scope of scientific inquiry, suggesting that human consciousness might possess properties beyond our traditional understanding. Parapsychology explores telepathic experiences and their possible explanations, ranging from mirror neurons and brainwave synchronization to speculative links with quantum mechanics. Quantum consciousness theories, such as the idea of entanglement and nonlocality, propose that minds may be interconnected in ways that allow for direct mental communication.

In conclusion, scientific investigations into message encounters reveal the vast, largely uncharted territories of the mind. From the practical application of hypnosis to the philosophical questions posed by quantum consciousness, these studies push the boundaries of what is possible, inviting us to consider that the mind's potential extends far beyond known limits. While many questions remain unanswered, these explorations bring us closer to understanding how individuals perceive, interpret, and retain experiences that lie at the edge of human understanding.

As we move from examining the scientific frameworks behind perception, memory, and consciousness, we arrive at a compelling question: What messages are being conveyed in extraterrestrial encounters, and are there patterns or themes that recur across different cases? Studies on message content reveal intriguing consistencies, suggesting that these communications may share common threads despite cultural and personal differences. In this section, we will explore the themes that emerge most frequently in reported encounters—such as environmental warnings, spiritual guidance, and apocalyptic prophecies—and examine case studies that reinforce these patterns. Additionally, we'll draw comparisons with other extraordinary phenomena, like near-death experiences, mystical visions, and shamanic journeys, to identify parallels that may point to universal aspects of the human psyche or shared encounters with something beyond our current understanding. This exploration into patterns and commonalities offers a window into the possible intentions behind these messages and hints at the role they might play in our collective evolution.

Chapter 6: Unveiling the Message: Patterns and Commonalities in Extraterrestrial Communications

Throughout history, reports of extraterrestrial encounters have been marked by intriguing similarities, with recurring themes that span cultures, generations, and geographic locations. As we delve into the messages allegedly conveyed in these encounters, certain patterns emerge, suggesting an overarching narrative that may hold significance for humanity as a whole. From warnings of environmental degradation and impending disasters to calls for spiritual awakening and evolution, these themes seem to transcend individual experience and point toward shared, possibly universal, messages. This section will explore these recurring themes in detail, examining case studies that highlight their consistency and significance. We'll also compare these themes with those found in other extraordinary experiences—such as near-death experiences, mystical visions, and shamanic journeys—shedding light on the common threads that link these phenomena. By understanding these patterns and commonalities, we can begin to piece together a more cohesive picture of the intentions behind these messages and consider their potential implications for the future of humanity.

Unveiling Universal Warnings: Identifying Common Themes in Extraterrestrial Messages

Across numerous extraterrestrial encounters, from personal testimonies to widely reported events, we see a recurring pattern in the messages conveyed to humanity. These themes often center on crucial topics: environmental preservation, spiritual evolution, and sometimes, warnings of potential apocalyptic outcomes. Despite the diversity of witnesses, cultures, and contexts, these common themes resonate across encounters, suggesting an intentional narrative designed to alert, guide, or even transform our collective consciousness. This exploration delves into these central themes, examining the motivations behind them and their relevance to our modern world. Understanding these messages offers a profound opportunity to engage with what might be universal truths or essential insights for humanity's future.

Environmental Warnings: Messages to Protect Our Planet

One of the most prominent themes in extraterrestrial messages is the urgent call to address environmental degradation. Witnesses often report receiving warnings about the dire state of Earth's ecosystems and the need to halt destructive behaviors such as deforestation, pollution, and climate change. This theme is particularly striking, as it reflects concerns shared by many environmentalists and scientists, bridging the gap between human concerns and purported extraterrestrial interest in Earth's well-being.

Case Studies Highlighting Environmental Themes

Several well-documented encounters reflect a deep concern for Earth's environment, as conveyed by extraterrestrial beings. In these cases, individuals report receiving clear messages about humanity's responsibility to protect and preserve the planet.

- **Ariel School Encounter (1994)**: In this famous case, a group of schoolchildren in Ruwa, Zimbabwe, claimed to have encountered extraterrestrial beings who communicated, through mental impressions, warnings about environmental destruction. The children reported that the beings conveyed the idea that humanity's actions were causing harm to the planet, emphasizing the need for immediate change. This

encounter has become a powerful example of extraterrestrial messages urging environmental awareness.
- **Betty Andreasson's Abduction (1967)**: Betty Andreasson, an American woman who claimed to have experienced an abduction, reported that her extraterrestrial captors emphasized the importance of caring for Earth. She described being shown vivid images of environmental devastation, which she interpreted as a warning to humanity about the consequences of ecological neglect.

Motivations Behind Environmental Warnings

The repeated focus on environmental concerns in extraterrestrial messages raises questions about the motivations of these purported beings. Some theorists propose that extraterrestrials may see Earth as a unique biosphere that must be preserved, possibly due to its interconnected ecosystems or the value it holds within a larger cosmic framework. Others suggest that these beings may have observed similar patterns of environmental degradation on their own planets, which could explain their apparent desire to prevent humanity from following a similar path. Whatever the motivation, the consistent emphasis on environmental protection suggests that these messages are intended to inspire action and a sense of responsibility toward the planet.

Spiritual Messages: Calls for Human Evolution and Enlightenment

Another central theme in extraterrestrial encounters is the emphasis on spiritual development and the need for humanity to evolve beyond materialistic and divisive thinking. Many witnesses report receiving messages encouraging compassion, unity, and a deeper understanding of life's purpose. These spiritual messages often contain elements that parallel traditional religious or philosophical teachings, yet they are presented as universal truths rather than dogmatic beliefs.

Case Studies Reflecting Spiritual Themes

Several encounters have documented extraterrestrial messages that focus on humanity's spiritual evolution, with beings encouraging individuals to elevate their consciousness and embrace values that foster collective harmony.

- **The Contactee Movement (1950s)**: During the mid-20th century, a wave of "contactees"—individuals who claimed to have established communication with extraterrestrial beings—reported receiving messages that promoted peace, love, and spiritual growth. Figures like George Adamski and Howard Menger described their interactions with beings from other planets as inspirational, with extraterrestrials urging humans to adopt a more enlightened way of life.
- **Whitley Strieber's Communion**: In his book *Communion*, Whitley Strieber describes experiences with beings who seemed to emphasize his own spiritual awakening. Strieber's encounters highlight a theme of personal transformation, with the beings prompting him to confront his fears, embrace self-awareness, and consider the interconnectedness of all life.

Possible Interpretations of Spiritual Messages

The spiritual messages in extraterrestrial encounters may be interpreted as guidance for humanity to reach a higher level of consciousness. Some theorists suggest that these beings might have achieved advanced levels of spirituality themselves, prompting them to share insights that could help humanity evolve in similar ways. These messages often align with principles of compassion, mindfulness, and nonviolence, encouraging people to transcend fear-based reactions and adopt a more holistic perspective. In many cases, experiencers report feeling a profound sense of love or oneness after their encounters, which reinforces the possibility that these messages are intended to foster a collective shift in consciousness.

Apocalyptic Warnings: Visions of Catastrophe and the Urge to Prepare

A more ominous theme that emerges in some encounters is the warning of potential apocalyptic events. These messages often describe visions of natural disasters, war, or societal collapse, with the beings urging humanity to change course before it is too late. Apocalyptic messages are usually presented as conditional outcomes, suggesting that while disaster is possible, it can be avoided if humanity alters its behavior. Such warnings evoke both fear and urgency, as they paint a picture of the possible consequences of inaction.

Notable Cases Featuring Apocalyptic Themes

The presence of apocalyptic warnings in extraterrestrial encounters has been documented in several high-profile cases, each describing different aspects of potential catastrophe.

- **The Fatima Apparitions (1917)**: While traditionally viewed as a religious event, the Fatima apparitions contain elements that resemble extraterrestrial encounters. The children who witnessed the event reported receiving a series of visions, including scenes of war, suffering, and natural disasters, which were interpreted as warnings about humanity's future. Many researchers have since examined these apparitions as potential extraterrestrial communications, noting the similarities in message content.

- **Travis Walton's Abduction (1975)**: Travis Walton, who claimed to have been abducted by extraterrestrials, reported seeing scenes of devastation and environmental degradation. Though Walton's experience was traumatic, he described a sense of foreboding that suggested a need for humanity to confront these challenges or face severe consequences. His experience aligns with other abduction accounts that feature apocalyptic warnings.

Understanding the Intent Behind Apocalyptic Warnings

Apocalyptic warnings from extraterrestrial beings may reflect a desire to motivate humanity by presenting the stark consequences of continued destructive behavior. These messages often emphasize humanity's agency, suggesting that catastrophic events are avoidable if people make conscious, collective changes. Some researchers theorize that these warnings might be rooted in extraterrestrials' own experiences with societal collapse, while others view them as symbolic representations of the dangers inherent in unbridled technological advancement and environmental neglect. Regardless of the beings' motivations, the apocalyptic theme highlights a sense of urgency and a call for immediate action.

Common Threads in Extraterrestrial Messages

The recurring themes in extraterrestrial messages—environmental responsibility, spiritual evolution, and apocalyptic warnings—offer insights into the possible intentions behind these encounters. Environmental messages emphasize the importance of protecting Earth's ecosystems, reflecting both a concern for planetary health and a potential awareness of humanity's impact on the environment. Spiritual messages encourage individuals to embrace compassion, unity, and a higher purpose, suggesting that extraterrestrials may recognize the need for humanity's evolution on a consciousness level. Apocalyptic warnings, while ominous, often serve as conditional alerts, pushing humanity to make changes that could prevent devastating outcomes.

Together, these themes form a cohesive narrative that speaks to humanity's responsibilities, potential, and vulnerabilities. Whether viewed as cautionary tales or as guides for transformation, these messages urge humanity to act with mindfulness, care, and a sense of interconnectedness. By identifying and analyzing these common threads, we gain a better understanding of the messages conveyed in extraterrestrial encounters and the possible reasons these themes persist. In studying these patterns, we are reminded of the potential impact these messages may have on our collective future, inspiring us to consider what role they might play in shaping our path forward.

The Consistency of Extraterrestrial Messages: Case Studies Unveiling Shared Themes Across Encounters

Reports of extraterrestrial encounters often convey messages that appear strikingly consistent across diverse cases, time periods, and geographic regions. Many of these experiences, whether they involve direct encounters, abductions, or contact with unidentified flying objects (UFOs), reveal themes that resonate deeply with the current global concerns of humanity. Environmental warnings, calls for spiritual growth, and apocalyptic forewarnings are frequently reported by individuals who claim to have communicated with extraterrestrial beings. This consistency suggests that these messages may not merely be the products of personal imagination or cultural influence but instead could point to a universal narrative—one that emphasizes both the urgent and hopeful aspects of human potential and responsibility.

In this exploration, we delve into specific case studies that showcase the consistent themes across extraterrestrial messages, using examples that vividly bring to life the patterns and concerns that appear in these experiences. Each case offers insights into the recurring themes, providing a foundation for understanding the broader implications of these messages for humanity's evolution.

Ariel School Encounter: A Vision of Environmental Stewardship

One of the most compelling cases illustrating the theme of environmental concern is the Ariel School encounter, which took place in Ruwa, Zimbabwe, in 1994. In this incident, a group of schoolchildren between the ages of six and twelve reported seeing a landed UFO near their school and encountering beings who telepathically communicated warnings about humanity's environmental impact. These messages, though conveyed to children, emphasized the importance of protecting the Earth and avoiding destructive behaviors that could lead to severe consequences.

The Power of Innocence: Why the Message to Children Matters

The choice of children as recipients of this message is notable; as young individuals with limited exposure to global issues, the children's accounts were seen as less likely to be influenced by pre-existing environmental concerns. This encounter underscores the potential importance of environmental stewardship and may reflect an extraterrestrial attempt to reach humanity through witnesses who would not easily dismiss or reinterpret the experience. The Ariel School encounter is a powerful example of how environmental themes in extraterrestrial messages resonate across encounters, offering a simple yet profound message: to protect our planet.

The Betty Andreasson Abduction: A Personal Call for Environmental and Spiritual Awareness

In the United States, Betty Andreasson's reported abduction in 1967 introduced themes of both environmental and spiritual awareness, woven together in a narrative that profoundly impacted her life. During her experience, Andreasson reported seeing vivid images of environmental devastation, including scenes of polluted rivers and desolate landscapes. Her extraterrestrial captors allegedly conveyed that these images represented humanity's future if it continued on its destructive path.

A Dual Message: Environmental and Spiritual Responsibility

What sets Andreasson's encounter apart is the blend of environmental and spiritual themes. In addition to warning her about the dangers of ecological neglect, her captors emphasized the importance of inner transformation and spiritual growth. Andreasson interpreted this dual message as a call for humans to reconnect with both the Earth and their higher selves. This encounter exemplifies how messages from extraterrestrial beings often encompass both physical and spiritual dimensions, suggesting that humanity's relationship with the environment is inherently linked to its collective consciousness and moral evolution.

The Contactee Movement: Messages of Peace and Spiritual Unity

During the 1950s and 1960s, the Contactee Movement emerged as a unique phenomenon in which individuals claimed to have established communication with extraterrestrial beings, often referred to as "Space Brothers." Key figures within this movement, such as George Adamski and Howard Menger, reported receiving messages that emphasized peace, love, and spiritual unity. These messages typically depicted extraterrestrials as benevolent beings who sought to guide humanity toward a more harmonious future.

Peace and Unity in the Atomic Age

The timing of the Contactee Movement, during the height of the Cold War and the beginning of the nuclear arms race, is significant. The messages received by contactees often reflected concerns about nuclear weapons and the potential for global conflict. Adamski, for example, described his extraterrestrial contacts as warning humanity about the dangers of nuclear war and encouraging a shift toward peaceful coexistence. The consistency of these messages across multiple contactees, coupled with their timing, suggests a deliberate effort to influence human behavior in a period of heightened tension and existential risk. This case highlights the way extraterrestrial messages can echo humanity's own anxieties, amplifying calls for change in moments of crisis.

The Fatima Apparitions: A Religious Encounter with Potential Extraterrestrial Overtones

The 1917 Fatima apparitions, traditionally viewed as a religious event, have been interpreted by some researchers as a possible extraterrestrial encounter. In this case, three young shepherd children in Portugal reported receiving a series of visions and messages from an entity they identified as the Virgin Mary. These messages included predictions of global suffering, urging humanity to repent and turn toward peace to avoid catastrophe. While this event is generally framed within a religious context, the themes align closely with those observed in modern extraterrestrial encounters, particularly the apocalyptic warnings and the call for spiritual reflection.

A Cross-Interpretation of Extraterrestrial and Religious Themes

The Fatima apparitions are intriguing because they combine traditional religious elements with themes found in extraterrestrial messages, such as apocalyptic predictions and the call for spiritual transformation. Some researchers suggest that this encounter might reflect a broader phenomenon in which divine or otherworldly messages are conveyed to humanity in forms that resonate with the cultural beliefs of the time. In Fatima, the appearance of the "Miracle of the Sun" and the emphasis on peace and repentance mirror the motifs present in other extraterrestrial messages, bridging religious and extraterrestrial themes. This case underscores how encounters can take on different interpretations based on cultural context, while maintaining consistent underlying messages.

The Travis Walton Abduction: Apocalyptic Warnings of Earth's Future

The Travis Walton abduction, which occurred in 1975 in Arizona, presents another case where environmental and apocalyptic warnings were central to the experience. Walton, a logger, reported being abducted by extraterrestrial beings and witnessing scenes of environmental degradation and human suffering. These images, which he described as deeply disturbing, appeared to reflect a future in which humanity failed to address critical issues such as deforestation, pollution, and social discord.

A Darker Perspective: Urgency and the Consequences of Inaction

Unlike some encounters that focus on hope and guidance, Walton's experience carried a darker tone, emphasizing the potential consequences of human inaction. This encounter resonates with other apocalyptic-themed messages, suggesting that the beings may have sought to instill a sense of urgency in Walton, motivating him to share these warnings with a wider audience. The Walton case contributes to the pattern of apocalyptic themes, as it aligns with other accounts where extraterrestrials appear to be cautioning humanity against self-destructive behaviors.

Identifying Consistent Themes Across Encounters

Examining these case studies reveals a remarkable consistency in the themes communicated through extraterrestrial encounters, regardless of location, time period, or cultural background. Messages centered on environmental responsibility, spiritual evolution, and apocalyptic warnings appear across diverse cases, suggesting that these are not isolated narratives but part of a broader, unified message aimed at guiding humanity toward a sustainable and enlightened future.

The Ariel School encounter highlights environmental concerns, underscoring the need for ecological preservation through the innocence of children. Betty Andreasson's experience blends spiritual and environmental themes, reinforcing the idea that humanity's outer actions are deeply tied to its inner consciousness. The Contactee Movement's focus on peace and unity during a time of nuclear tension reflects the extraterrestrials' awareness of human affairs and their attempt to steer humanity away from conflict. The Fatima apparitions illustrate how these themes can cross religious and cultural boundaries, while the Walton abduction offers a sobering vision of potential consequences should humanity fail to heed these messages.

In sum, these cases collectively point to a narrative that emphasizes the urgent need for change in human attitudes and behaviors. Whether these messages are truly extraterrestrial or reflections of collective subconscious anxieties, the themes they present challenge humanity to reflect on its values, responsibilities, and the interconnectedness of all life. The consistency of these messages, across different contexts, invites us to consider the possibility of a shared wisdom beyond our world, encouraging humanity to awaken to its potential and live in harmony with itself and the Earth.

Beyond Encounters: Comparative Insights from Near-Death Experiences and Mystical Visions

The messages received in extraterrestrial encounters often touch on themes that extend beyond earthly concerns and seem to resonate on a deeply personal and spiritual level. Interestingly, these themes—environmental responsibility, spiritual enlightenment, and apocalyptic warnings—are also commonly found in experiences traditionally classified as near-death experiences (NDEs) and mystical visions. By examining these phenomena side-by-side, we can gain a deeper understanding of whether the messages reported in extraterrestrial encounters share a universal quality, rooted perhaps not only in our fears and hopes but also in a potential connection to a shared consciousness or otherworldly insight.

In this comparative analysis, we'll explore the striking parallels between extraterrestrial messages and the transformative insights reported by individuals who have undergone NDEs or mystical visions. Understanding these similarities provides context for interpreting extraterrestrial messages and raises profound questions about the nature of consciousness, spirituality, and the boundaries of human experience.

Near-Death Experiences and Extraterrestrial Messages: Parallel Encounters with Other Realms

Near-death experiences have been documented across cultures and time periods, typically involving individuals who have come close to death and subsequently report visions or encounters with beings of light, deceased loved ones, or divine figures. Often, these experiences impart a profound sense of purpose and a heightened awareness of life's interconnectedness. Likewise, extraterrestrial encounters frequently convey transformative messages that emphasize humanity's unity, the sanctity of life, and the urgency of protecting our planet.

Shared Themes of Unity and Purpose

In both NDEs and extraterrestrial messages, there is a recurring emphasis on unity—an idea that all life is interconnected, and that actions taken by one individual or group inevitably impact the whole. NDE experiencers often describe an overwhelming sense of oneness with the universe, along with a newfound awareness that they are intrinsically connected to all beings. This mirrors the messages reported by many who have encountered extraterrestrial beings, who claim to receive messages encouraging humanity to transcend divisive behavior and adopt a more harmonious approach to life.

- **Example from NDE Literature**: Dr. Eben Alexander, a neurosurgeon who experienced an NDE, described a sense of profound interconnectedness and a feeling that love is the fundamental force binding the universe. His experience included visions of otherworldly realms, where beings communicated telepathically with him, echoing the method and content of many extraterrestrial messages.

- **Extraterrestrial Encounter Parallel**: In the case of Travis Walton, who reported being abducted in 1975, his experience also involved a vision of life's interconnectedness. Although his experience was traumatic, he described a growing awareness of how humanity's actions affect the Earth, highlighting an underlying message about collective responsibility and environmental stewardship.

The Transformative Power of Encounters

Both NDEs and extraterrestrial encounters often have life-altering impacts on those who experience them. Many NDE experiencers return to life with a new sense of purpose, often abandoning materialistic or self-centered pursuits in favor of more meaningful, compassionate lives. Similarly, individuals who report encounters with extraterrestrials often undergo a profound shift in values, describing a heightened sense of responsibility to the Earth, humanity, and even the universe as a whole.

- **Personal Transformation in NDEs**: After her near-death experience, author and researcher P.M.H. Atwater described feeling an intense connection to nature and humanity, which led her to study and document NDEs. Like many other experiencers, she felt an urgent need to share the insights she had received, emphasizing compassion, empathy, and care for the planet.
- **Parallel in Extraterrestrial Messages**: The Contactee Movement of the 1950s is filled with accounts of people who claimed that their encounters with extraterrestrials led them to adopt pacifist, environmentally conscious lifestyles. Figures like George Van Tassel and Howard Menger reported that their "Space Brothers" encouraged peace, unity, and respect for the Earth, sparking a movement rooted in spiritual and ecological awareness.

Mystical Visions and Extraterrestrial Encounters: Glimpses into Hidden Realities

Mystical visions, often occurring during meditation, fasting, or moments of crisis, have long been described by mystics and seekers as encounters with higher beings, divine realms, or otherworldly landscapes. In many ways, these experiences bear a striking resemblance to certain extraterrestrial encounters, with both phenomena often featuring vivid imagery, telepathic communication, and themes of cosmic revelation.

Encounters with Higher Beings

A common element in mystical visions is the appearance of beings who impart wisdom, guidance, or warnings. These figures are frequently perceived as benevolent entities, and the visions often include messages of love, peace, and unity. Similarly, individuals who report extraterrestrial encounters often describe beings who communicate telepathically and convey similar messages about universal love, interconnectedness, and the importance of moral evolution.

- **Mystical Vision Example**: The 13th-century Sufi mystic Jalaluddin Rumi described his mystical experiences as encounters with "beings of light" who guided him toward a deeper understanding of love and unity. Rumi's visions became the foundation for his poetry, which emphasized the idea that love and compassion connect all creation.
- **Extraterrestrial Encounter Parallel**: In the famous 1961 abduction of Betty and Barney Hill, Betty described telepathic communication with an extraterrestrial entity who conveyed a sense of calm and emphasized the importance of mutual respect and understanding. Although the Hill encounter was initially traumatic, Betty's experience included

moments that parallel the guidance and peace described by mystics like Rumi.

Revelations of Cosmic Order and Humanity's Place in the Universe

In both mystical visions and extraterrestrial messages, there is often a focus on revealing humanity's place within a greater cosmic order. Mystics report experiences that give them insight into the structure of the universe and the laws that govern existence. Similarly, extraterrestrial encounters often include messages that suggest humans are part of a larger, interconnected cosmic community, and that our actions have consequences that extend beyond our world.

- **Mystical Insight Example**: The German mystic Hildegard of Bingen, a 12th-century abbess and visionary, described a series of visions that she interpreted as revelations of divine wisdom. Her writings speak of cosmic harmony, balance, and humanity's role as both caretaker and participant in this divine order.
- **Extraterrestrial Encounter Parallel**: During the 1980s, Whitley Strieber reported encounters with beings who communicated insights about consciousness, the interconnectedness of all life, and the idea that humanity must evolve to achieve its full potential. Strieber's accounts often reflect the themes found in mystical visions, with a focus on humanity's role within a larger, universal framework.

Comparing Experiences: Are These Encounters Expressions of the Same Source?

The striking similarities between NDEs, mystical visions, and extraterrestrial messages raise fascinating questions about the origins of these experiences. Do they represent unique interpretations of a shared phenomenon, or are they different expressions of a universal truth? Some researchers suggest that these experiences may stem from a single source of wisdom or knowledge, accessible to individuals in altered states of consciousness or moments of heightened awareness. This theory posits that whether framed as extraterrestrial encounters, NDEs, or mystical visions, these experiences are all attempts by the human mind to interpret a profound, otherworldly reality.

The Role of Cultural Interpretation

One explanation for the differences in framing—extraterrestrial versus religious or mystical—may be cultural context. For example, a medieval mystic who experiences a vision of radiant beings might interpret them as angels, while a modern individual may view similar beings as extraterrestrial. This suggests that while the core themes and messages remain consistent, the interpretation may vary depending on societal influences and personal belief systems.

A Universal Call for Change?

Another interpretation is that these experiences reflect a universal call for humanity to recognize its responsibility and potential. Whether perceived as messages from extraterrestrials, glimpses of the afterlife, or divine revelations, these experiences frequently convey the same essential messages: protect the Earth, nurture spiritual growth, and recognize the interconnectedness of all life. These similarities imply that humanity is being repeatedly called to awaken to its higher purpose, regardless of the source of these messages.

Parallels in Purpose and Message Across Phenomena

The analysis of near-death experiences, mystical visions, and extraterrestrial messages reveals a pattern of strikingly similar themes that transcend individual interpretation. Each of these phenomena, though often framed differently, conveys messages about unity, environmental stewardship, and the need for spiritual growth. From the environmental warnings communicated in extraterrestrial encounters to the profound sense of interconnectedness experienced in NDEs, these messages appear to be part of a broader, universal narrative that speaks to humanity's highest ideals and most pressing responsibilities.

The overlap between these experiences suggests that the distinctions we draw between different types of encounters may be less important than the underlying messages themselves. Whether experienced through extraterrestrial contact, near-death episodes, or mystical visions, these messages encourage humanity to adopt a holistic worldview that emphasizes compassion, unity, and awareness. This comparative analysis offers a lens through which we can view these encounters as expressions of a shared wisdom, calling us to embrace a path of transformation that honors both the Earth and the broader cosmos. Ultimately, the consistency of these themes across phenomena invites us to consider that perhaps these messages are not simply personal or cultural constructs, but part of a universal truth waiting for humanity to recognize and embody.

Journeys Beyond the Known: Similarities Between Extraterrestrial Messages, Shamanic Journeys, and Prophetic Dreams

Throughout history, shamanic journeys and prophetic dreams have provided humans with access to hidden realms, sources of wisdom, and insights into both personal and collective futures. Remarkably, many of the themes encountered in these altered states of consciousness align closely with the messages reported in extraterrestrial encounters. Like contactees and abductees, shamans and prophets describe interactions with otherworldly beings, the transmission of messages about humanity's responsibilities, and visions of both hope and warning. Examining these parallels can illuminate how these experiences intersect and may even originate from a shared source of knowledge, accessed through diverse paths of perception.

In this exploration, we'll delve into how extraterrestrial messages mirror the transformative journeys of shamans, as well as the apocalyptic and visionary content often found in prophetic dreams. Together, these comparisons can provide a richer understanding of extraterrestrial messages by revealing their connections to humanity's ancient methods of spiritual exploration and insight.

Shamanic Journeys: Encounters with Otherworldly Beings and Hidden Knowledge

For thousands of years, shamans have journeyed into realms beyond ordinary reality to interact with spirits, gain knowledge, and offer guidance to their communities. These journeys, induced by trance, fasting, or the use of sacred plants, bring the shaman into direct contact with beings described as guides, ancestors, or celestial figures. In many ways, these beings share remarkable similarities with those reported in extraterrestrial encounters, often conveying messages about the environment, humanity's behavior, and the importance of harmony.

The Shamanic Journey as a Gateway to Cosmic Wisdom

In shamanic traditions, the shaman enters altered states to access a realm that many believe is separate from yet interwoven with our own. Here, they communicate with entities who often emphasize humanity's role in maintaining balance and respect for the natural world. The beings encountered by shamans frequently provide insights into ecological harmony, encouraging sustainable practices and a reverence for all life—themes commonly found in extraterrestrial messages.

- **Example of Shamanic Wisdom**: Among the Siberian Tungus people, shamans are known to commune with spirits who provide knowledge on the cycles of nature and offer guidance for sustaining the community's resources. These messages reflect a belief that humanity's role is not to dominate but to coexist within the larger ecosystem, a theme that aligns closely with the environmental warnings frequently reported in extraterrestrial encounters.
- **Parallel with Extraterrestrial Messages**: Many individuals who claim to have experienced extraterrestrial contact describe receiving telepathic messages that emphasize environmental stewardship and the interconnectedness of all life. The Ariel School encounter, in which children received warnings about the Earth's environmental health, mirrors the messages imparted to shamans about the importance of ecological balance. Both the shamanic and extraterrestrial narratives emphasize that humanity's survival depends on respecting and preserving nature.

Beings of Light and the Shamanic Experience

In shamanic journeys, the beings encountered are often described as luminous or radiant, sometimes appearing as animals, ancestors, or celestial figures. These beings are not always viewed as "gods" but as intermediaries or teachers, sharing knowledge that the shaman is responsible for conveying back to their community. This imagery—of luminous beings imparting wisdom—strongly parallels reports of extraterrestrial encounters, where beings of light are often reported as conveying messages of unity, peace, and responsibility.

- **Luminous Beings in Shamanism**: In Amazonian shamanism, ayahuasca ceremonies often bring participants into contact with radiant beings who share visions and insights. Many report that these beings, though awe-inspiring, encourage humility, peace, and unity. This experience of encountering higher, compassionate entities closely mirrors the experiences of those who report meeting extraterrestrials who impart similar messages.
- **Extraterrestrial Encounter Parallel**: In the accounts of Betty Andreasson, a well-known contactee, extraterrestrial beings of light communicated messages of spiritual and environmental wisdom. Her descriptions of these beings as luminous and profoundly wise echo the imagery and themes of shamanic journeys, where the beings encountered are often seen as custodians of knowledge and guides for humanity.

Prophetic Dreams: Visions of Apocalyptic Warnings and Collective Evolution

Prophetic dreams are another domain where humanity has historically accessed foreknowledge and warnings. In these dreams, individuals often witness future events, apocalyptic scenarios, or symbols of transformation. Many cultures interpret prophetic dreams as messages from divine sources or as glimpses into the collective psyche. Similarly, extraterrestrial messages often contain visions of global disasters or calls for spiritual awakening, suggesting that both phenomena serve as warnings or guidance for humanity's evolution.

Apocalyptic Warnings in Prophetic Dreams and Extraterrestrial Messages

One of the most consistent themes in prophetic dreams is the vision of a catastrophic event or apocalyptic future. In many accounts, dreamers are shown symbols of destruction—floods, fires, wars, or other calamities—that imply consequences for humanity's actions. This theme is mirrored in many extraterrestrial encounters, where individuals report seeing images of environmental devastation or societal collapse, often coupled with a plea for humanity to change its ways.

- **Historical Example of Prophetic Dreams**: In the Bible, the prophet Daniel's dreams often included visions of future kingdoms rising and falling, emphasizing both transformation and catastrophe. These dreams were not merely personal; they were considered messages for entire societies. Similarly, ancient Egyptian and Greek cultures valued prophetic dreams for their perceived ability to reveal truths about collective fate.
- **Extraterrestrial Encounter Parallel**: During the 1960s and 1970s, contactees frequently reported receiving apocalyptic messages about nuclear war and environmental destruction. The beings reportedly showed them images of Earth in ruins, urging humanity to abandon war and adopt a peaceful path. These warnings closely resemble the imagery found in prophetic dreams, suggesting that the messages may stem from a shared intent to guide humanity away from potential disasters.

Visions of Transformation: Humanity's Evolution and Spiritual Awakening

Beyond catastrophe, prophetic dreams often contain symbols of rebirth, transformation, or the need for spiritual evolution.

These dreams can serve as calls to action, encouraging individuals to elevate their consciousness, embrace compassion, and fulfill a greater purpose. Extraterrestrial messages echo this theme, as many contactees describe receiving instructions for personal or collective growth, often urging humanity to overcome materialistic or divisive tendencies.

- **Spiritual Awakening in Prophetic Dreams**: Indigenous cultures, such as the Hopi, hold a tradition of prophetic dreaming where visions are seen as messages from the ancestors or spirit world. These dreams often urge the dreamer to maintain spiritual purity and guide their community toward a harmonious future. In Hopi prophecy, the imagery of the "Great Purification" reflects a transformation that humanity must undergo, embodying a call to consciousness similar to that found in extraterrestrial encounters.
- **Parallel in Extraterrestrial Encounters**: Whitley Strieber's experiences with beings that communicated messages of spiritual awakening reflect a parallel with prophetic dreams. Strieber reported receiving guidance on transcending fear, understanding the interconnectedness of all life, and achieving a higher level of consciousness. This message of transformation is not limited to extraterrestrial encounters; it resonates with the messages received by prophets, mystics, and dreamers across history.

The Intersection of Extraterrestrial Messages, Shamanic Journeys, and Prophetic Dreams

The similarities between shamanic journeys, prophetic dreams, and extraterrestrial messages suggest that these experiences may tap into a shared source of wisdom that transcends the boundaries of ordinary perception. Whether accessed through altered states, dreams, or encounters with otherworldly beings, these experiences convey messages that speak to humanity's responsibilities, potential, and relationship with the cosmos. The recurring themes—environmental stewardship, spiritual evolution, and apocalyptic warnings—point to a universal concern for humanity's future and a desire to guide us toward a harmonious, sustainable existence.

A Unified Narrative of Responsibility and Potential

What makes these parallels especially compelling is the consistency of the messages across vastly different contexts and cultures. Shamanic journeys, prophetic dreams, and extraterrestrial encounters all encourage individuals and societies to confront their actions and recognize their interconnectedness with the world around them. This suggests that these experiences may not merely be isolated phenomena but parts of a larger, cohesive narrative that underscores humanity's potential to evolve and the consequences of failing to do so.

Are These Encounters Expressions of the Same Source?

The consistency of the messages raises intriguing questions about their origins. Some researchers suggest that these encounters may all be different interpretations of a universal, archetypal experience, where the mind interacts with a dimension of reality that conveys insights about human existence. Alternatively, these encounters could be distinct but parallel manifestations, each revealing aspects of a shared wisdom or truth intended to help humanity navigate its path forward.

A Convergence of Wisdom Across Realms

The comparison of extraterrestrial messages with shamanic journeys and prophetic dreams reveals a profound convergence of themes and messages that transcend cultural and experiential boundaries. In all three realms, individuals report encounters with beings who emphasize the importance of environmental responsibility, spiritual growth, and preparation for possible apocalyptic outcomes. These similarities suggest that humanity may be receiving a consistent call to awaken to its role in the universe, regardless of the specific method of communication.

By examining these experiences together, we gain insight into the possibility of a shared source of knowledge that invites humanity to explore its potential and responsibilities. Whether interpreted as divine guidance, ancestral wisdom, or extraterrestrial communication, these messages collectively urge us to respect the natural world, pursue inner growth, and recognize our interconnectedness. As we continue to seek understanding, the parallels across these phenomena remind us that the wisdom offered in these encounters may be part of a larger tapestry, woven to guide humanity toward a more conscious and sustainable existence.

Patterns and Commonalities in Extraterrestrial Messages

Exploring the themes within extraterrestrial encounters reveals a consistent narrative that aligns with humanity's most profound concerns: environmental stewardship, spiritual evolution, and apocalyptic foresight. Across encounters, whether they involve direct telepathic communication or visions imparted during abductions, these themes serve as a collective message aimed at guiding humanity toward greater awareness and responsibility. Case studies, from the Ariel School incident to Betty Andreasson's abduction, show that the same themes emerge repeatedly across diverse settings and witnesses, strengthening the argument that these messages are universal rather than random.

When we compare extraterrestrial messages with other transformative experiences—such as near-death experiences, mystical visions, shamanic journeys, and prophetic dreams—a remarkable similarity emerges. All these phenomena convey messages that urge individuals and societies to act with greater respect for life, pursue spiritual understanding, and prepare for possible cataclysmic events if change is not embraced. These parallels suggest a common thread of wisdom that transcends the boundaries of individual perception, cultural interpretation, and the specific nature of the encounter.

In sum, the recurring themes and their presence across various experiences hint at a possible shared source of guidance, one that may be urging humanity to evolve, protect the Earth, and seek deeper truths about our place in the universe. These insights provide a compelling foundation for further study, inviting us to consider whether the messages delivered in these encounters might serve as a catalyst for humanity's transformation and awakening. By recognizing and integrating these patterns, we open the door to a more conscious and responsible relationship with our world and perhaps with other realms beyond our own.

Having explored the recurring themes and patterns within extraterrestrial messages, along with their fascinating parallels in near-death experiences, shamanic journeys, and prophetic dreams, we now turn to deeper questions about the nature and purpose behind these encounters. While the messages themselves suggest a call for environmental preservation, spiritual growth, and awareness of potential dangers, the intentions of the beings conveying these messages remain open to interpretation. Are these entities benevolent guides, warning and preparing us for challenges ahead? Or do they have hidden motives, potentially influencing humanity for their own purposes?

In this next section, we will examine various interpretations and theories about extraterrestrial intentions. From evaluating whether these messages indicate a benevolent or malevolent intent, to exploring how extraterrestrial civilizations might perceive humanity and why they might intervene, we delve into the complex motives behind these interactions. Additionally, we'll consider speculative theories on how extraterrestrials might play a role in humanity's evolution and what future interactions could entail based on the content of these messages. This analysis provides a foundation for understanding not only the messages themselves but also the broader implications of these potential interstellar relationships.

Chapter 7: Unmasking the Motives: Interpretations and Theories about Extraterrestrial Intentions

As we shift from understanding the content of extraterrestrial messages to analyzing the intentions behind them, we confront one of the most compelling questions in the study of these encounters: Why would extraterrestrial beings seek to communicate with humanity? The themes present in these messages—warnings of environmental damage, calls for spiritual growth, and hints of apocalyptic consequences—suggest that these beings may have insights into our civilization's challenges and potential. But are they acting as benevolent guides, genuinely invested in our well-being, or do they have other, perhaps less altruistic, motives?

In this section, we explore the wide spectrum of theories regarding extraterrestrial intentions. We examine whether these beings approach humanity with a benevolent or malevolent agenda and consider the ethical implications of their potential influence on human affairs. We also delve into hypotheses about how extraterrestrial civilizations might perceive humanity and why they might choose to interact with us. Finally, we look to speculative theories about a possible extraterrestrial role in human evolution and consider the future of these interactions based on the current patterns in their messages. This exploration invites us to consider not only what extraterrestrials are saying but also why they might be saying it, opening a window into the broader significance of these encounters for humanity's future.

Friend or Foe? Understanding Benevolent and Malevolent Intent in Extraterrestrial Messages

Encounters with extraterrestrial beings and their accompanying messages provoke a profound question: What motivates these entities to communicate with humanity, and are their intentions truly in our best interest? The recurring themes of environmental concern, spiritual growth, and apocalyptic warning suggest a possible guiding or advisory role, but interpretations of these messages vary widely. Some theorists posit that extraterrestrials act as benevolent guides, aiming to help humanity avoid disaster and embrace a higher path. Others caution that these beings may have ulterior motives, potentially influencing humanity for reasons that serve their own agendas rather than ours. Exploring the potential intentions behind extraterrestrial messages requires a careful analysis of both the themes conveyed and the context in which they are received, as well as insights from historical, ethical, and psychological perspectives.

In this exploration, we delve into interpretations of extraterrestrial intentions, considering the evidence and arguments on both sides of the benevolent-malevolent spectrum. By examining the possible motivations behind these messages, we gain a clearer understanding of whether extraterrestrial encounters represent genuine concern for human welfare or if they serve purposes that may be less aligned with our interests.

Benevolent Beings: Extraterrestrials as Guides and Guardians

One of the most common interpretations of extraterrestrial intentions is that these beings operate from a place of compassion and concern for humanity. Proponents of this view suggest that extraterrestrial messages often reflect a higher moral and ethical standard, one that emphasizes peace, harmony, and the protection of our planet. According to this perspective, extraterrestrials might be more advanced civilizations, both technologically and spiritually, who are reaching out to us as mentors, offering guidance to help humanity avoid self-destruction.

Environmental and Spiritual Warnings as Acts of Altruism

The consistency of messages focused on environmental protection and spiritual evolution lends credence to the view that extraterrestrials may be benevolent guides. Messages that encourage humanity to avoid war, protect nature, and embrace compassion and unity appear to serve the interests of humanity itself, suggesting an altruistic intent. In cases like the Ariel School encounter, children reported that extraterrestrial beings conveyed a sense of urgency about humanity's destructive impact on the planet, a message that would appear to benefit Earth's ecosystem and future generations.

- **Case Study: The Contactee Movement**: During the 1950s, contactees such as George Adamski and Howard Menger reported receiving messages from extraterrestrials who warned of the dangers of nuclear weapons and the potential for global catastrophe. These extraterrestrials, described as "Space Brothers," were said to emphasize peace and love, encouraging humanity to adopt a more enlightened path. The themes in these messages align closely with humanitarian values, suggesting that extraterrestrials might be motivated by a desire to prevent humanity from making the same mistakes they may have witnessed in their own civilizations.

The Ethical Implications of Non-Interference

Many proponents of the benevolent theory argue that extraterrestrials are careful not to interfere overtly with human affairs, adhering to an ethical principle of non-interference akin to the "Prime Directive" concept popularized in science fiction. This notion implies that while extraterrestrials may observe and offer subtle guidance, they respect humanity's free will and right to self-determination, intervening only when absolutely necessary. The fact that extraterrestrials appear to communicate primarily through telepathic messages or subtle visual displays, rather than direct physical intervention, might indicate a careful adherence to ethical boundaries.

- **Insights from Cultural Anthropology**: Anthropological studies of isolated human tribes provide a useful analogy. In these cases, modern societies often refrain from interfering with isolated tribes, understanding that doing so could disrupt their cultural integrity and autonomy. Similarly, extraterrestrials may view humanity as an evolving civilization, one that they hope will reach its full potential without overt intervention. By providing warnings and guidance without dictating outcomes, they may be encouraging us to make conscious choices rather than imposing their own solutions.

Darker Motives: Are Extraterrestrials Acting in Self-Interest?

While the benevolent interpretation is reassuring, other theories suggest that extraterrestrials may be motivated by self-interest or even malevolence. Skeptics argue that the themes in extraterrestrial messages could serve to manipulate human beliefs, behaviors, and attitudes in ways that ultimately benefit extraterrestrials rather than humanity. From this perspective, the messages of environmental or apocalyptic warnings might be designed to induce fear, compliance, or dependence on these beings, potentially setting the stage for extraterrestrials to exert control or exploit resources.

Manipulation Through Apocalyptic Messaging

One of the arguments for a potentially malevolent intent is the focus on apocalyptic warnings in some extraterrestrial messages. While these warnings might appear altruistic, skeptics point out that fear is a powerful tool for controlling behavior. By presenting visions of catastrophe, extraterrestrials could be instilling a sense of urgency and dependency, positioning themselves as saviors or protectors. This dynamic could make humanity more receptive to extraterrestrial influence or intervention, subtly altering our decision-making processes to align with extraterrestrial goals.

- **Example of Apocalyptic Influence**: The Fatima Apparitions, though traditionally viewed as a religious event, included warnings of global calamity if humanity did not repent. Some researchers interpret this encounter as a possible extraterrestrial event, noting the psychological impact that apocalyptic visions have on individuals and communities. In this light, apocalyptic warnings may be a strategy to instill obedience or fear, encouraging humanity to look to extraterrestrials as guides or protectors without critically questioning their motives.

Ethical Concerns and Potential Exploitation

From an ethical standpoint, extraterrestrials might have motivations that are not immediately visible or fully understood by humanity. Just as human interactions with other species or ecosystems can sometimes lead to unintended exploitation, extraterrestrials might approach humanity with their own set of goals that may not necessarily align with our best interests. This perspective suggests that extraterrestrials may view humanity as a resource, whether for biological, intellectual, or planetary reasons, rather than as equals deserving of respect.

- **Historical Analogy of Colonialism**: Some theorists draw parallels between extraterrestrial interactions and historical instances of colonialism, where more technologically advanced societies exploited less advanced cultures for resources, labor, or strategic advantage. The messages encouraging environmental conservation and spiritual evolution could thus be a means to preserve humanity as a stable and manageable civilization, one that extraterrestrials could potentially exploit or manipulate for their own benefit.

A Nuanced Middle Ground: Extraterrestrials as Observers and Experimenters

An alternative interpretation of extraterrestrial intentions considers the possibility that extraterrestrials are neither purely benevolent nor malevolent but rather neutral observers or experimenters. According to this theory, extraterrestrials might be studying humanity as part of a larger interest in planetary civilizations, intervening only minimally to gather data or observe humanity's response to certain stimuli. In this view, extraterrestrials could be likened to scientists studying an evolving species, intrigued by our development but largely dispassionate about the outcomes.

Extraterrestrial Messages as Part of a Scientific Experiment

If extraterrestrials are observers, then their messages might be intended to test human responses rather than to direct human behavior. By issuing messages that emphasize environmental, spiritual, or apocalyptic themes, extraterrestrials may be gauging humanity's collective reaction, analyzing how different cultures and individuals interpret and act on these warnings. This approach would align with an interest in understanding humanity's psychological, social, and spiritual development, rather than explicitly guiding or protecting us.

- **Possible Evidence of Experimentation**: The varied nature of extraterrestrial messages, ranging from peaceful encouragement to terrifying warnings, suggests that extraterrestrials might be observing a spectrum of human responses. Some individuals report positive, transformative experiences, while others experience fear or trauma. This range of experiences could imply that extraterrestrials are exploring human resilience, adaptability, and interpretive diversity, rather than providing a clear directive.

A Spectrum of Intentions in Extraterrestrial Encounters

In analyzing the motivations behind extraterrestrial messages, we encounter a spectrum of interpretations that range from altruistic guidance to potential manipulation. The benevolent perspective suggests that extraterrestrials seek to aid humanity in overcoming environmental, spiritual, and existential challenges, acting as mentors who respect our autonomy and wish to see us thrive. Alternatively, the more skeptical view posits that extraterrestrials may have self-serving motives, possibly manipulating humanity through fear or dependency to advance their own goals. Finally, the neutral observer hypothesis suggests that extraterrestrials may simply be studying humanity, issuing messages as part of a broader experiment in understanding developing civilizations.

Each perspective offers valuable insights into the possible nature of extraterrestrial intentions. While we may never fully understand the true motivations behind these messages, exploring these theories helps us to remain discerning, thoughtful, and open-minded as we continue to interpret the messages conveyed in extraterrestrial encounters. Whether these beings are guides, manipulators, or observers, the messages they deliver urge us to consider our responsibilities and potential as a species, challenging us to reflect deeply on humanity's role in the cosmos.

Intervention or Independence? Ethical Perspectives on Extraterrestrial Interference with Humanity

When considering extraterrestrial encounters and the messages conveyed within them, we are faced with a critical ethical question: Should extraterrestrial beings interfere in human affairs, and if so, under what circumstances? Interference, whether perceived as guidance, influence, or outright control, raises a host of ethical dilemmas. These questions are complex, not only because they address our fundamental rights to autonomy and self-determination but also because they force us to confront our assumptions about what extraterrestrial civilizations might consider ethical or responsible behavior.

In this exploration, we'll delve into various ethical perspectives on extraterrestrial interference, considering arguments for and against involvement, drawing on analogies from human history, and examining the potential motivations behind such actions. Through this analysis, we aim to uncover how ethical perspectives on extraterrestrial intervention shape our understanding of their intentions and impact our approach to encounters with otherworldly intelligences.

A Duty to Guide? The Case for Benevolent Intervention

From one perspective, interference by extraterrestrial beings could be viewed as a morally responsible action if it serves to protect humanity from its own self-destructive tendencies. This perspective suggests that an advanced civilization, having likely overcome similar challenges, might feel a moral obligation to assist younger civilizations in navigating existential risks. Much like a mentor who intervenes to prevent a mentee from making irreversible mistakes, extraterrestrials might see interference as a compassionate act aimed at ensuring humanity's survival.

Moral Responsibility to Prevent Harm

Proponents of benevolent interference argue that if extraterrestrials possess knowledge of potential catastrophic outcomes—such as environmental collapse, nuclear war, or social disintegration—they might be ethically compelled to intervene. This perspective aligns with utilitarian principles, which emphasize actions that maximize well-being and minimize harm. If extraterrestrials have foreseen potential disasters that humanity is unaware of or unwilling to address, withholding this knowledge could be considered morally negligent, especially if their guidance could save lives or preserve the planet.

- **Analogy with Human Intervention**: In cases of human societies, intervention is sometimes justified when there is a severe threat to life, such as humanitarian aid in war-torn regions or disaster relief. Likewise, extraterrestrials might perceive that the gravity of certain human actions justifies interference, especially if it prevents harm to humanity or even to the planet as a whole.

Intervention as a Form of Compassionate Mentorship

Some theorists propose that extraterrestrials might view humanity as a developing civilization that requires guidance to reach its full potential. According to this view, interference is not about control but about offering gentle nudges that allow humanity to grow in a way that aligns with universal principles of peace, unity, and sustainability. Rather than imposing direct solutions, extraterrestrials might choose to subtly influence humanity's trajectory through messages, encouraging people to reach conclusions and solutions independently.

- **Insights from Child Development**: Just as parents guide children, offering support and wisdom without overtly controlling them, extraterrestrials might feel ethically obligated to intervene in ways that promote growth while respecting humanity's autonomy. This approach would represent a delicate balance between influence and respect, allowing humanity to evolve on its own path while benefiting from occasional guidance.

The Right to Self-Determination: Arguments Against Interference

Conversely, many argue that extraterrestrials have no right to interfere with humanity's development, regardless of their intentions. According to this perspective, interference undermines humanity's autonomy and violates the principle of self-determination—a core tenet in ethics and international relations. Humanity, as an independent civilization, must be allowed to navigate its challenges without external influence, even if those challenges result in failure or self-destruction.

Respect for Autonomy and Cultural Integrity

The concept of autonomy emphasizes the importance of making one's own choices, even if those choices carry risks. This view argues that if extraterrestrials interfere, they may inadvertently disrupt humanity's natural development, imposing their values, technologies, or philosophies in a way that erodes cultural integrity. Just as respecting the autonomy of individuals or nations prevents coercive control, respecting humanity's right to choose its path safeguards the cultural diversity and uniqueness of our species.

- **Historical Example of Cultural Erosion**: The history of colonialism demonstrates the dangers of intervention by more advanced societies, which often led to the suppression of indigenous cultures and traditions. If extraterrestrials interfere, they may unintentionally initiate a form of cultural erosion, causing humanity to lose its unique identity and sense of agency. Advocates of non-interference argue that humanity must be allowed to make its own choices, for better or worse, to preserve its cultural integrity.

The Potential for Dependency

Another ethical concern related to interference is the potential for dependency. If extraterrestrials intervene frequently, humanity might begin to rely on their guidance, losing its capacity for independent problem-solving and growth. This dependency could undermine humanity's resilience, creativity, and adaptability, qualities that are essential for any civilization's long-term survival. By interfering, extraterrestrials might inadvertently weaken humanity's ability to face challenges on its own, creating a lasting dependence on external intervention.

- **Psychological Insights on Dependency**: In psychology, it is well understood that over-reliance on external support can inhibit personal development. Similarly, if humanity begins to expect or depend on extraterrestrial guidance, our civilization may lose the drive to solve its own problems. Opponents of intervention argue that humanity's independence must be protected, ensuring that we retain the skills, knowledge, and agency required to confront future challenges without external help.

The Ethical Dilemma of Selective Intervention

An intriguing perspective considers the possibility that extraterrestrials may adopt a policy of selective intervention, stepping in only under specific circumstances while maintaining a general stance of non-interference. Selective intervention, however, raises its own ethical questions, as extraterrestrials would need to determine criteria for involvement—such as the severity of the threat, humanity's readiness, or potential for harm. This approach resembles an ethical compromise, but it risks creating ethical gray areas where the line between acceptable guidance and undue influence becomes blurred.

Defining the Limits of Intervention

One ethical challenge with selective intervention lies in defining its boundaries. For instance, extraterrestrials might choose to intervene to prevent global nuclear war but refrain from offering solutions to social, economic, or environmental issues. While this approach minimizes direct interference, it also assumes that extraterrestrials are capable of predicting which actions are in humanity's best interest. This presumption could lead to unintended consequences if extraterrestrials misinterpret humanity's needs or overestimate the impact of their guidance.

- **Comparing to Human Intervention**: Selective intervention mirrors certain practices in human international relations, where intervention is carefully limited to prevent direct influence over another nation's internal affairs. However, such selective involvement can still be contentious, as it implies a power dynamic where one party claims the authority to intervene only when deemed "necessary." In the context of extraterrestrial involvement, this approach could be seen as presumptuous, suggesting that extraterrestrials are better equipped to understand humanity's needs than humanity itself.

Balancing Non-Interference with Humanitarian Concerns

Selective intervention also presents a moral dilemma, as extraterrestrials would need to balance their commitment to non-interference with a potential ethical obligation to prevent suffering. For example, if they possess knowledge that could alleviate a natural disaster or cure a disease, withholding this information could be viewed as morally questionable. Yet, sharing such knowledge might inadvertently disrupt human progress, alter cultural values, or interfere with scientific discovery.

- **Ethics in Humanitarian Aid**: The debate around providing aid in certain crisis situations reflects this ethical tension. While aid can alleviate suffering, it can also disrupt local development and create dependency. Similarly, extraterrestrials would need to weigh the benefits of intervention against the long-term impact on humanity's development, finding a balance that respects our autonomy while honoring a commitment to reduce harm.

The Principle of the "Prime Directive": Lessons from Fictional Ethics

Science fiction, particularly the "Prime Directive" in *Star Trek*, explores the ethical implications of non-interference with developing civilizations. According to this directive, more advanced civilizations are morally obligated to avoid interfering with the natural progression of less advanced societies. This principle, though fictional, resonates with the ethical concerns surrounding extraterrestrial involvement, as it encapsulates the ideal of allowing civilizations to reach their potential without external influence.

The "Prime Directive" as a Model for Ethical Interference

The Prime Directive suggests that extraterrestrials, if they adhere to a similar philosophy, would refrain from influencing humanity's development unless absolutely necessary. This stance respects humanity's right to self-determination and avoids the pitfalls of dependency, cultural erosion, and unintended consequences. However, the Prime Directive also poses challenging ethical questions: Is there ever a scenario where breaking this non-interference rule is justified? If humanity's survival is at stake, would it be ethical for extraterrestrials to uphold a rigid policy of non-interference?

- **Hypothetical Scenario**: Imagine an extraterrestrial civilization witnessing humanity's approach to environmental collapse. If they follow a "Prime Directive" approach, they might choose to observe rather than intervene, hoping humanity finds its own solutions. However, if their knowledge could prevent disaster, would it be ethical to withhold it? This dilemma reflects the tension between respecting autonomy and exercising moral responsibility, a core challenge in the ethics of extraterrestrial interference.

Navigating the Ethics of Extraterrestrial Interference

The question of extraterrestrial interference raises profound ethical considerations, as it requires balancing respect for humanity's autonomy with potential responsibilities to prevent harm. Proponents of benevolent interference argue that extraterrestrials, if motivated by compassion, might have a duty to assist humanity in avoiding catastrophic outcomes, serving as guides and mentors. Others, however, caution that interference risks eroding cultural integrity and creating dependency, arguing that humanity must face its challenges independently to preserve its autonomy.

Selective intervention offers a potential middle ground but introduces ethical ambiguities regarding when and how extraterrestrials might justify their involvement. The "Prime Directive" model, while fictional, provides a valuable framework for considering the implications of non-interference, emphasizing the importance of allowing civilizations to evolve without undue influence. Ultimately, understanding the ethics of extraterrestrial interference sheds light on the broader intentions behind these encounters, prompting us to reflect on humanity's right to self-determination, the potential role of advanced civilizations as mentors, and the moral complexities of interstellar relationships.

Through Alien Eyes: Hypotheses on How Extraterrestrial Civilizations Perceive Humanity

Imagining how an extraterrestrial civilization might view humanity is a journey into speculation informed by anthropology, psychology, and our own history of intercultural encounters. While humanity's understanding of extraterrestrials is limited to the accounts and experiences of those who claim contact, these stories offer intriguing hints about how extraterrestrials might perceive our species. Are we seen as fledgling beings on the cusp of cosmic maturity, or as a potential danger to the galactic community? Understanding how extraterrestrials might view us could provide insights into their motivations, their reasons for communication, and their possible reluctance to reveal themselves openly.

In this exploration, we'll investigate various hypotheses about how extraterrestrial beings might interpret humanity. These perspectives range from seeing us as a developing civilization worthy of guidance to perceiving us as potentially dangerous or chaotic. By analyzing each theory in depth, we gain a clearer sense of why extraterrestrials may interact with humanity in subtle or controlled ways, shaping the broader context of interstellar relationships.

Humanity as an Evolving Species: A Young Civilization in Cosmic Development

One hypothesis suggests that extraterrestrials see humanity as a young, evolving species—similar to how an adult views a child or how one might observe an emerging society. From this perspective, extraterrestrials may recognize our intellectual and technological achievements but also see us as lacking the maturity or wisdom that comes with advanced development. Just as humans observe the behavior of adolescents, waiting to see how they will grow and develop, extraterrestrials may view humanity as a civilization in the early stages of its cosmic journey.

A Civilization with Potential: Waiting for Maturity

If extraterrestrials perceive us as an evolving civilization, they may see potential for growth and possibly even for eventual membership in a larger galactic community. Just as educators or mentors might watch over the progress of promising students, extraterrestrials may observe humanity's evolution with a cautious optimism. However, this perspective would also likely come with a measure of patience, as extraterrestrials wait for humanity to reach certain ethical, spiritual, or intellectual milestones before initiating direct, open contact.

- **Example of Maturity Indicators**: Certain achievements, such as uniting to solve global issues, advancing scientific understanding without endangering other species, and developing sustainable ways of life, might be indicators that extraterrestrials look for in determining our readiness. As humanity works toward goals like environmental conservation, space exploration, and peaceful collaboration, extraterrestrials might interpret these as signs that humanity is progressing toward a state of maturity.

Observing with Caution and Encouragement

From this perspective, extraterrestrials might send messages aimed at encouraging specific types of growth, such as spiritual evolution, environmental responsibility, or cooperation. Like a guiding hand that remains unseen, extraterrestrials might subtly shape humanity's development without overwhelming our autonomy or influencing our natural trajectory. This approach would allow humanity to grow and evolve at its own pace, with extraterrestrials observing as we overcome challenges and develop a greater understanding of ourselves and our universe.

- **Parallel with Human Mentorship**: In education or mentorship, guides often encourage students by providing resources or planting ideas rather than overtly controlling their development. Extraterrestrials might employ a similar strategy, subtly guiding humanity through messages that encourage reflection and growth, all while allowing us to make our own decisions.

Humanity as a Potential Threat: Assessing the Risks of Contact

Another perspective suggests that extraterrestrials may view humanity with caution, seeing us as unpredictable or potentially dangerous. Given humanity's history of violence, environmental exploitation, and conflict, extraterrestrials might interpret our civilization as one that could pose a risk if we were to gain advanced technology or knowledge without first achieving ethical maturity. This perspective frames humanity as a species that, if left unchecked, could disrupt cosmic harmony or even pose a direct threat to other civilizations.

Concerns About Aggression and Environmental Neglect

Humanity's propensity for war and our often-destructive relationship with the environment might be cause for extraterrestrial concern. From their perspective, a civilization that engages in frequent conflict and disregards the health of its own planet may lack the values necessary to interact responsibly with other species. This could lead extraterrestrials to adopt a policy of limited contact, observing humanity but refraining from offering technology or knowledge that might be misused.

- **Historical Analogy with Nuclear Weapons**: When humans developed nuclear weapons, it marked a turning point in our ability to destroy not only ourselves but also other species and the planet itself. Extraterrestrials might see our handling of such power as a litmus test for ethical restraint and responsibility, recognizing that a civilization willing to harness such destructive force may not yet be ready for advanced extraterrestrial interaction.

Containment Through Limited Engagement

If extraterrestrials perceive humanity as a potential threat, their interactions might be deliberately limited to observation or carefully controlled messages. This approach would allow extraterrestrials to monitor our progress without providing any tools or knowledge that could be weaponized. In this sense, their actions could be seen as a form of containment, reducing the risk of a volatile species causing harm to itself or others while still allowing for observation and potential guidance.

- **Parallel with Conservation Practices**: In conservation biology, scientists often study potentially dangerous species in a way that limits direct interaction, ensuring that both the observers and the observed are kept safe. Extraterrestrials might take a similar stance, carefully controlling their interactions with humanity to prevent accidental harm or escalated risks.

Humanity as a Scientific Curiosity: An Experiment in Evolution and Consciousness

Some theorists propose that extraterrestrials might view humanity not as a potential ally or threat but as a scientific curiosity. From this perspective, humanity could be of interest as a case study in the development of consciousness, cultural evolution, or biological adaptation. Extraterrestrials with a strong interest in scientific exploration might observe humanity in the same way that humans study animals or plants, documenting our behaviors, beliefs, and social structures.

A Civilization Under Observation: The Anthropological Approach

If extraterrestrials view humanity as a subject for study, their approach might resemble the anthropological techniques humans use to study other cultures or species. They might observe from a distance, occasionally engaging in controlled encounters, and refrain from intervening directly in human affairs. In this view, extraterrestrials would approach humanity with curiosity and fascination, carefully recording our progress while withholding information or resources that could affect the integrity of their study.

- **Ethical Observational Approach**: In anthropology, researchers often adopt a stance of non-interference to preserve the authenticity of their observations. Extraterrestrials might see humanity in a similar light, choosing to observe rather than interact to gain unaltered insights into human behavior, culture, and evolution. This approach would prioritize the purity of data over influencing our development, allowing extraterrestrials to understand humanity's natural trajectory.

Studying Consciousness and Emotional Development

Another possible reason for extraterrestrial observation could be to understand consciousness, particularly human emotions, creativity, and spirituality. Humanity's psychological complexity might intrigue extraterrestrials, especially if they themselves possess different cognitive structures. Studying human responses to concepts like love, fear, and spirituality could offer extraterrestrials a unique glimpse into aspects of existence that may differ from their own.

- **Parallel in Animal Behavior Studies**: Just as humans study animal behavior to understand evolutionary adaptations and social dynamics, extraterrestrials might study humanity to gain insights into the nature of consciousness. Observing human culture, art, and emotional responses could provide extraterrestrials with valuable information about the diversity of intelligent life and its expressions across the universe.

Humanity as a Future Ally or Contributor to Galactic Civilization

A more optimistic hypothesis suggests that extraterrestrials might see humanity as a potential ally or contributor to a larger cosmic community. From this perspective, humanity's unique strengths—such as adaptability, creativity, and resilience—could eventually make us valuable members of an interstellar alliance or coalition. Extraterrestrials with this perspective might interact with us in a way that promotes unity, wisdom, and cooperation, preparing humanity for the eventual possibility of formal contact.

Encouraging Development for Future Cooperation

Extraterrestrials who view humanity as a future ally might engage in a long-term strategy to prepare us for membership in a galactic community. This might involve sending messages that encourage peace, environmental stewardship, and scientific advancement, helping us reach a state where we are capable of collaborating responsibly with other species. By cultivating a sense of cosmic citizenship within humanity, extraterrestrials could be laying the groundwork for future alliances.

- **Example of Diplomatic Patience**: Just as nations engage in diplomacy with developing countries, extraterrestrials might exercise patience in their interactions with humanity, gradually fostering qualities that align with interstellar values. In this view, extraterrestrial messages would serve as guidance for humanity's evolution, preparing us for eventual integration rather than immediate interaction.

Seeing Humanity's Strengths as a Unique Contribution

Extraterrestrials might also recognize strengths in humanity that could contribute to the broader galactic community. Qualities such as creativity, resilience, and diversity may be seen as assets that, once refined, could enrich the larger cosmic fabric. This perspective would not only acknowledge humanity's current limitations but also affirm our potential to grow into a civilization capable of offering unique insights, innovations, and cultural contributions.

- **Parallel with Cultural Exchange**: In cultural exchanges between nations, each culture brings distinct perspectives and strengths. Similarly, extraterrestrials might value humanity's contributions, understanding that our diversity and adaptability could add depth and richness to a galactic society. This perspective emphasizes cooperation, recognizing that each civilization has something valuable to offer.

Extraterrestrial Perspectives on Humanity— From Observation to Collaboration

Understanding how extraterrestrials might view humanity provides valuable insights into their intentions and approaches to contact. If extraterrestrials see us as a young, evolving civilization, they may offer guidance in subtle ways, encouraging growth without overwhelming our autonomy. Alternatively, if they perceive humanity as a potential threat, their interactions may be limited to observation and containment, avoiding actions that could empower us before we are ready for responsible use of advanced knowledge.

In another view, extraterrestrials might see humanity as a scientific curiosity, studying us as a unique example of biological, cultural, and spiritual development. This perspective would prioritize observation over intervention, focusing on the insights humanity offers into the nature of consciousness and adaptation. Finally, some may see humanity as a future ally or contributor, encouraging our development with the hope that one day we will join a larger galactic community.

Each of these perspectives provides a different lens through which we can interpret extraterrestrial messages and interactions, challenging us to reflect on humanity's place in the cosmos and our readiness to engage with other civilizations. Whether as observers, guides, or potential allies, extraterrestrials' perspectives on humanity reveal as much about their values and ethics as they do about our own civilization's journey.

Why They Might Intervene: Exploring Extraterrestrial Motivations for Guiding Humanity

As we ponder the reasons extraterrestrial civilizations might intervene in or guide humanity, a range of intriguing possibilities emerges. Would an advanced species see us as a younger sibling needing mentorship, a civilization at risk of self-destruction, or a species with unique qualities worth nurturing? Understanding why extraterrestrials might choose to involve themselves in our affairs requires us to consider their potential values, interests, and ethical frameworks, as well as the long-term impact their guidance could have on humanity.

In this exploration, we'll consider various motivations for extraterrestrial intervention, from a sense of moral duty and shared responsibility to practical concerns about human behavior and its cosmic implications. Through this lens, we gain a deeper understanding of how an advanced civilization might approach the question of intervention and the circumstances that could prompt them to engage with humanity.

A Sense of Cosmic Stewardship: Intervening as a Moral Duty

One hypothesis suggests that extraterrestrials might feel a sense of moral responsibility to help younger or less advanced civilizations. If extraterrestrials have achieved high levels of spiritual, ethical, or intellectual maturity, they may view the guidance of other species as part of a broader duty to promote harmony and balance in the universe. This sense of stewardship would position them not as conquerors or manipulators, but as guardians who seek to protect and uplift.

The Guardian's Code: Advanced Morality and Ethical Obligation

For a civilization that values peace, compassion, and the preservation of life, it may be ethically unacceptable to remain passive while humanity struggles with challenges like environmental degradation, social division, and nuclear proliferation. From this perspective, extraterrestrials might feel an intrinsic duty to offer guidance, much like a seasoned traveler who warns newcomers of potential dangers on a shared journey. Their involvement could be motivated by a desire to prevent avoidable suffering or to help humanity reach its full potential without unnecessary hardship.

- **Parallel with Humanitarian Aid**: Just as humanitarian organizations intervene in crisis situations to prevent human suffering, extraterrestrials might see themselves as ethically obligated to assist a species on the brink of self-destruction. Their guidance might aim to address immediate crises, offering humanity insights or technologies that could help us overcome obstacles and preserve our civilization for future generations.

Cultivating Harmony Across the Cosmos

This sense of stewardship could be rooted in a broader commitment to cosmic harmony. If extraterrestrials view the universe as an interconnected system in which all beings play a role, then the development of any civilization affects the whole. From this viewpoint, humanity's choices, successes, and failures would impact other worlds or species, giving extraterrestrials a vested interest in helping us achieve harmony and avoid actions that could disturb universal balance.

- **Insights from Ecological Ethics**: In ecology, the interconnectedness of species within ecosystems underscores the importance of maintaining balance. Similarly, extraterrestrials might see themselves as guardians of cosmic balance, intervening to ensure that developing civilizations like humanity do not disrupt the delicate equilibrium of the cosmos.

Preventing Self-Destruction: Intervention to Preserve Potential

Another motivation for extraterrestrial intervention could be a concern that humanity is at risk of self-destruction. With our ability to harness nuclear power, manipulate the environment, and engage in large-scale warfare, humanity possesses the means to bring about its own extinction. For extraterrestrials who recognize our potential, intervening to prevent self-destruction may be less about interference and more about preserving a valuable asset—our capacity for creativity, resilience, and innovation.

Preventing Tragic Loss of Potential

If extraterrestrials see value in humanity's unique qualities—our creativity, adaptability, or capacity for emotional depth—they may consider it a tragic loss if humanity were to self-destruct. This perspective would frame intervention as a means of safeguarding the development of a civilization that could one day contribute meaningfully to the galactic community. Such guidance would focus on encouraging humanity to embrace sustainable practices, peaceful coexistence, and ethical use of technology.

- **Example of Preservation Efforts**: Just as conservationists strive to preserve endangered species, extraterrestrials may see themselves as conservationists for civilizations, stepping in to prevent humanity from becoming an existential casualty. Their guidance might include warnings against environmental destruction, reckless scientific pursuits, or conflicts that risk catastrophic outcomes.

Encouraging Responsible Use of Advanced Technology

With humanity's rapid advancements in technology comes the risk of misuse, whether through weaponization, environmental exploitation, or other harmful applications. Extraterrestrials, recognizing the dangers of unregulated technological growth, might feel compelled to offer advice on responsible practices. They could provide insights into the ethical use of technology, advocating for a balance between innovation and caution, much like how regulatory bodies oversee scientific research to ensure safety and ethical integrity.

- **Parallel with Nuclear Non-Proliferation Efforts**: In international relations, agreements and treaties are established to prevent the spread of nuclear weapons, recognizing the destructive potential of such technology. Extraterrestrials might adopt a similar approach, encouraging humanity to limit its use of technologies that could threaten planetary stability or cosmic harmony, thus preserving both human and universal welfare.

Fostering Unity and Peace: Humanity as a Future Ally

Some theorists suggest that extraterrestrials may view humanity as a potential ally or contributor to a larger interstellar community. From this perspective, humanity's diversity, resilience, and creativity could make us valuable members of a cosmic alliance once we reach a level of maturity that aligns with peaceful and cooperative values. Extraterrestrials might intervene with the intention of fostering qualities that would make humanity a stable, harmonious member of this broader collective.

Encouraging a Global Sense of Unity

One of the primary obstacles to humanity's participation in an interstellar community is internal division—conflict, nationalism, and inequality create barriers to global cooperation. Extraterrestrials who see potential in humanity might seek to inspire a sense of unity, encouraging us to overcome these divisions in favor of collaboration and shared purpose. Through messages that emphasize the interconnectedness of life and the importance of mutual respect, extraterrestrials could be gently guiding us toward the values needed to join a peaceful cosmic society.

- **Analogy with Diplomatic Alliances**: In human diplomacy, countries work toward alliances that require them to align on shared values, goals, and practices. Extraterrestrials might adopt a similar approach with humanity, aiming to cultivate the characteristics of unity, tolerance, and responsibility that would facilitate our future role in an interstellar coalition.

Preparing Humanity for Cosmic Citizenship

If extraterrestrials are preparing humanity to take its place in a larger galactic society, their guidance might include insights into cosmic ethics, environmental stewardship, and technological restraint. This "training" would enable humanity to develop the wisdom, compassion, and self-discipline required to interact responsibly with other civilizations. By encouraging us to address our flaws and enhance our strengths, extraterrestrials could be paving the way for humanity to contribute meaningfully to an interstellar alliance.

- **Comparison with Peace Corps Initiatives**: Just as the Peace Corps supports developing nations by helping them build the skills and infrastructure needed for independence and collaboration, extraterrestrials might see humanity as a civilization in need of "galactic education." Their intervention could be akin to mentorship, preparing us to participate fully in a cosmic society rather than enforcing their values upon us.

Exploring New Realms of Consciousness: A Broader Vision of Reality

Some propose that extraterrestrials might intervene not to control or mentor humanity but to invite us to explore new dimensions of consciousness and existence. From this perspective, their guidance is less about practical concerns and more about helping humanity realize its full spiritual and cognitive potential. Extraterrestrials who see humanity as spiritually or intellectually underdeveloped might seek to expand our understanding of reality, introducing concepts that encourage exploration beyond the physical and material.

Opening the Doors of Perception

If extraterrestrials have reached advanced levels of consciousness, they might view humanity's current state as limited, bound by material concerns and unaware of deeper, more expansive truths. By encouraging practices like meditation, telepathy, or heightened environmental awareness, extraterrestrials might be inviting us to explore non-physical realms and elevate our understanding of life, reality, and consciousness.

- **Parallel with Mystical Teachings**: Just as mystics and spiritual teachers in human history have sought to guide others toward enlightenment, extraterrestrials may be attempting to catalyze an "awakening" within humanity. Their messages might emphasize themes of unity, interconnectedness, and the potential for transcending ordinary perception, encouraging us to embrace a broader vision of reality.

Expanding Human Consciousness for Future Collaboration

Extraterrestrials who engage in such guidance may be motivated by the belief that consciousness expansion is essential for humanity's evolution and, by extension, our readiness to join an interstellar community. By promoting awareness beyond physical limitations, they might be preparing us for interactions that involve telepathic communication, shared understanding, and experiences that defy current scientific paradigms. This approach would align with the idea that humanity's growth is not solely technological or social but also deeply connected to our inner awareness and spiritual development.

- **Insights from Psychedelic Research**: Studies in consciousness-altering substances, which reveal new dimensions of perception and interconnectedness, offer a parallel. Extraterrestrials might be providing us with tools or ideas that facilitate a "cosmic awakening," allowing humanity to expand its consciousness in preparation for more profound, multidimensional interactions with other civilizations.

Diverse Motivations for Extraterrestrial Intervention

Understanding why extraterrestrials might choose to intervene or guide humanity provides a rich framework for interpreting the messages and interactions reported in encounters. The motivations behind such guidance may stem from a deep sense of cosmic stewardship, with extraterrestrials feeling morally obligated to assist a civilization at risk of self-destruction. Alternatively, they may recognize humanity's potential as a future ally, fostering unity and collaboration to prepare us for interstellar citizenship. Others might see humanity as an emerging consciousness in need of expansion, guiding us toward a greater understanding of our own spiritual and intellectual capacities.

Each of these motivations reflects a different aspect of extraterrestrial interaction, whether as guardians, mentors, or awakeners. By exploring these diverse perspectives, we gain insight not only into why extraterrestrials might reach out to humanity but also into the possible values and philosophies that shape their actions. This understanding challenges us to reflect on our own development, offering both a mirror for self-evaluation and a potential pathway toward harmonious coexistence with other forms of intelligent life in the cosmos.

Shaping Destiny: Speculative Theories on Extraterrestrials' Role in Humanity's Evolution

The idea that extraterrestrials may play an active role in humanity's evolution offers a fascinating lens through which to consider our past, present, and future. Speculative theories about extraterrestrials influencing human development touch upon diverse fields—biology, spirituality, technology, and culture. These theories propose that extraterrestrials might not only be passive observers but also active participants, subtly or overtly guiding humanity's evolution to help us reach a specific destiny or potential. Whether through genetic manipulation, technological inspiration, or spiritual awakening, the idea that extraterrestrials have a hand in shaping humanity challenges our understanding of evolution and raises questions about our collective future.

In exploring these theories, we consider various potential roles extraterrestrials might play, from ancient architects influencing our biological development to cosmic mentors steering our spiritual growth. Each hypothesis offers unique insights into how extraterrestrials might view humanity and the reasons they might choose to intervene at critical junctures in our history.

The Ancient Architects: Theories of Genetic Influence and Hybridization

One prominent theory suggests that extraterrestrials have influenced human evolution on a genetic level, either through direct manipulation or by initiating hybridization programs. According to this hypothesis, humanity's evolutionary leaps—such as increased brain capacity, the development of language, or even the emergence of spirituality—may have been catalyzed by extraterrestrial intervention. Proponents of this theory argue that certain aspects of human biology are difficult to explain solely through natural evolution and that extraterrestrials may have intervened to accelerate our development or infuse specific traits into our DNA.

Genetic Manipulation as a Catalyst for Intelligence

Some theorists suggest that extraterrestrials may have altered early human DNA to enhance our intelligence, creativity, and adaptability. This hypothesis posits that humanity's capacity for abstract thought, technological innovation, and complex social structures may not have evolved naturally but was instead "seeded" by extraterrestrials to prepare humanity for a particular role or potential in the cosmos. By altering our genetic makeup, extraterrestrials could have set humanity on a path that diverges from that of other Earth species, pushing us toward the level of sentience and consciousness that might one day allow us to interact with other advanced civilizations.

- **Ancient Evidence in Mythology**: Many ancient myths and religious texts allude to gods or beings from the sky interacting with humans, sometimes creating hybrid offspring. Stories from the Sumerian, Egyptian, and Greek traditions describe gods who shape or guide humanity, often involving human-divine interbreeding. Proponents of the genetic influence theory argue that these myths could be ancient accounts of extraterrestrial interventions, recorded as best as early humans could interpret them.

Hybridization Programs and Modern Abduction Narratives

The theory of hybridization also appears in modern alien abduction stories, where individuals report encounters in which extraterrestrials perform medical procedures, sometimes involving reproductive manipulation. These accounts suggest that extraterrestrials may be engaged in an ongoing project to create hybrids, blending human and alien traits. Advocates of this theory believe that hybrid beings could be part of a long-term plan to introduce new capabilities into the human gene pool or to create a species that can bridge the gap between human and extraterrestrial understanding.

- **Case Study: The Hybridization Hypothesis**: Some abductees claim that extraterrestrials have expressed interest in human emotions, creativity, and adaptability. From this perspective, extraterrestrials might be attempting to incorporate these traits into a new hybrid species, blending the intellectual and physical strengths of both species. Such hybrids could serve as ambassadors between worlds, possessing both human empathy and extraterrestrial knowledge.

Extraterrestrials as Technological Catalysts: Seeding Innovation and Progress

Another theory proposes that extraterrestrials have intervened in human evolution by providing technological inspiration or even directly imparting knowledge. Throughout history, certain technological breakthroughs—whether in ancient engineering, medicine, or modern physics—have led humanity to rapid advancements. This theory suggests that these leaps in knowledge may have been influenced by extraterrestrials who selectively introduced concepts to stimulate human progress.

Ancient Civilizations and Possible Extraterrestrial Influence

Supporters of this theory often point to ancient structures like the Pyramids of Giza, Stonehenge, and the Nazca Lines, arguing that the technology and engineering required for these structures might have come from extraterrestrial sources. These monumental projects are not only impressive feats of architecture but also suggest an advanced understanding of astronomy, mathematics, and engineering. Extraterrestrials, according to this view, might have influenced or assisted ancient civilizations, guiding them in creating these structures as markers of cosmic alignment or as messages for future generations.

- **Symbolism and Mathematics in Ancient Monuments**: The precision with which ancient monuments align with celestial bodies has led some researchers to theorize that extraterrestrials shared knowledge of the cosmos with early humans. This transmission of astronomical knowledge could have served to encourage humanity's curiosity, stimulating an interest in science, the stars, and the broader universe.

Modern Technological Leaps and Extraterrestrial Guidance

In more recent times, proponents of this theory speculate that extraterrestrials may have had a hand in some of humanity's key technological breakthroughs, from atomic energy to computing and space travel. While no direct evidence supports this claim, anecdotal accounts and unusual scientific discoveries occasionally fuel the idea that extraterrestrials might subtly inspire or nudge humanity toward new technologies. This could be part of a long-term plan to prepare humanity for eventual integration into an interstellar network.

- **Example of Technological Inspiration**: Some theorists speculate that the rapid advancement of technology in the 20th century—from the development of nuclear energy to the rise of digital technology—could be attributed to extraterrestrial guidance or influence. By introducing technologies in controlled ways, extraterrestrials might be encouraging humanity's progress, helping us develop the tools we need to survive, communicate, and one day travel the cosmos.

Spiritual Evolution: Catalyzing a Shift in Human Consciousness

Beyond biological and technological influence, some theorists suggest that extraterrestrials are playing a role in humanity's spiritual evolution. This theory posits that extraterrestrials are not merely concerned with humanity's physical or intellectual growth but are deeply invested in our consciousness and spiritual development. Through telepathic communication, visions, or subtle guidance, extraterrestrials might be working to elevate humanity's awareness, encouraging us to transcend material concerns and embrace higher states of consciousness.

The Role of Extraterrestrials as Spiritual Mentors

Extraterrestrials who have reached advanced levels of spiritual understanding may view humanity as a species still entangled in fear, conflict, and ego. As spiritual mentors, they might offer guidance intended to inspire inner transformation, promoting values like compassion, unity, and harmony. This perspective aligns with messages received by contactees and abductees, many of whom report themes of spiritual enlightenment, interconnectedness, and the need to protect Earth.

- **Parallel with Human Mystics**: Throughout history, certain figures, such as prophets and mystics, have encouraged humanity to embrace compassion, peace, and wisdom. Extraterrestrials who act as spiritual mentors may be similar in intent, guiding humanity toward higher ideals and fostering a collective shift in consciousness that allows us to see beyond divisive boundaries.

Preparing Humanity for Multidimensional Realities

Some proponents of the spiritual evolution theory suggest that extraterrestrials are preparing humanity to perceive and interact with multidimensional aspects of reality. This preparation could involve stimulating psychic abilities, encouraging mindfulness and meditation practices, or teaching us to communicate telepathically. Extraterrestrials might believe that spiritual development is essential for humanity to truly understand and interact with other civilizations in a meaningful, non-violent way, fostering peace and cooperation on a cosmic scale.

- **Insights from Meditation and Consciousness Studies**: Just as meditation practices in various spiritual traditions aim to expand human awareness, extraterrestrials might be encouraging humanity to explore similar practices, viewing consciousness as the gateway to understanding the universe. By cultivating practices that elevate awareness, humanity could gradually develop the perceptual abilities needed for meaningful interactions with extraterrestrials.

Shaping Human Civilization as a Model for Others: The Earth Experiment

A final theory posits that Earth itself may be an experiment in evolution, diversity, and adaptability, watched over by extraterrestrials who are curious to see how humanity develops under specific conditions. In this scenario, extraterrestrials might be observing humanity as part of a larger study in the cosmic context, examining how diverse cultures, belief systems, and environmental factors shape a civilization. Humanity's journey would thus be seen as a model or case study for understanding broader principles of evolution and adaptation.

Humanity as a Case Study in Cultural and Ethical Evolution

If Earth is viewed as a social and ethical experiment, extraterrestrials might be interested in seeing how humanity manages its diversity, conflict, and environmental challenges. The many cultures, languages, and ideologies on Earth provide a rich tapestry for examining cooperation and discord, as well as the potential for harmony among diverse groups. Extraterrestrials might observe how humanity addresses issues like inequality, environmental preservation, and technological ethics, using these findings to inform their understanding of other civilizations.

- **Insights from Cross-Cultural Studies**: Anthropologists study various cultures to understand how different groups solve common problems. Similarly, extraterrestrials might observe Earth as a unique setting to explore the dynamics of cooperation, conflict, and innovation. By seeing how humanity navigates these issues, extraterrestrials could gain insights into the social evolution of other civilizations.

Testing Human Resilience and Adaptability

Extraterrestrials might also be examining humanity's resilience—our ability to adapt to environmental changes, innovate in the face of challenges, and rebuild after setbacks. This "Earth experiment" could be part of a larger cosmic investigation into the factors that allow civilizations to thrive or fail. By observing humanity's successes and failures, extraterrestrials might be gathering data to understand resilience, adaptability, and survival in complex environments.

- **Parallel with Ecosystem Studies**: In ecosystem studies, researchers observe how various species interact, adapt, and influence their environment. Extraterrestrials might take a similar approach, observing humanity's interaction with Earth to learn more about the factors that contribute to a civilization's long-term sustainability and resilience.

Extraterrestrial Influence on Humanity's Evolution—A Multifaceted Role

The idea that extraterrestrials may play a role in humanity's evolution opens up a rich array of possibilities, each reflecting a different aspect of our development. Theories suggest that extraterrestrials could influence our biological evolution through genetic manipulation, guiding us toward greater intelligence and creativity. They may act as technological catalysts, sparking innovations that drive human progress, or as spiritual mentors, encouraging a collective shift in consciousness. Alternatively, extraterrestrials might view Earth as a case study, observing our interactions, resilience, and adaptability to gain insights that could be applied to other civilizations.

Each of these roles—whether as architects, mentors, or observers—provides a different perspective on humanity's potential and the reasons extraterrestrials might intervene. Understanding these speculative theories helps us reflect on the broader questions of who we are, why we are here, and what role we may one day play in the vast cosmos. As we contemplate these possibilities, we gain not only insight into the motivations of extraterrestrial civilizations but also a deeper appreciation for the mystery and wonder of our own evolutionary journey.

Charting the Unknown: Hypothetical Future Interactions Between Extraterrestrials and Humanity

Speculating about future interactions with extraterrestrial beings based on current messages allows us to envision a wide range of potential relationships, each with profound implications for humanity. Extraterrestrial messages, often reported through encounters, abduction accounts, or alleged telepathic communications, frequently contain themes urging environmental preservation, spiritual growth, and the pursuit of peace. By examining the nature of these messages, we can anticipate hypothetical scenarios in which extraterrestrials may choose to engage with humanity more directly, shaping our future in transformative ways.

In this exploration, we will delve into the different paths these interactions might take, from gradual cultural exchanges to sudden transformative contact events. Through each scenario, we'll gain a nuanced understanding of how extraterrestrials might interact with humanity in the future, depending on how we respond to the guidance they appear to be providing.

The Path of Guidance: Gradual, Non-Invasive Mentorship

One possible future scenario for extraterrestrial interaction is a continuation of indirect guidance, with extraterrestrials providing subtle influence rather than overt presence. This model suggests that extraterrestrials may view humanity as a developing civilization that must mature at its own pace, free from the shock or disruption of direct contact. Instead, extraterrestrials might communicate through carefully crafted messages, promoting gradual changes in human consciousness, ethics, and social structure.

Shaping Global Consciousness through Indirect Influence

If extraterrestrials wish to guide humanity without directly intervening, they may choose to influence cultural evolution by encouraging a shift in values rather than providing explicit instruction. Messages encouraging unity, peace, and environmental awareness align with this approach, nudging humanity toward a path of responsible stewardship and cooperation. The extraterrestrials would remain observers, only intervening through indirect, symbolic means that inspire individuals and societies to reconsider their place within the larger cosmos.

- **Parallel with Religious and Philosophical Movements**: Just as religious and philosophical teachings throughout history have promoted ethical values without directly governing societies, extraterrestrial messages might serve as a philosophical influence on global consciousness. By prompting humanity to reflect on issues like climate change, war, and compassion, extraterrestrials could be seeding values that lay the groundwork for a more harmonious global society.

Spiritual Awakening and Cosmic Unity

Under this scenario, extraterrestrials may also aim to stimulate a spiritual awakening within humanity, gradually preparing us to see ourselves as part of a cosmic family. This approach could involve subtle nudges in the form of synchronicities, visionary experiences, or symbols that appear across different cultures. By guiding humanity toward greater spiritual unity, extraterrestrials might hope to prepare us for future interactions that would require openness, humility, and a broader sense of self beyond nationality or creed.

- **Comparison with the New Age Movement**: The rise of the New Age movement, with its focus on interconnectedness, meditation, and peace, could be seen as a parallel to this scenario. Just as New Age philosophies encourage individuals to embrace unity and compassion, extraterrestrials may be guiding humanity toward a similar ethos, creating a foundation for future engagement based on shared spiritual values.

Cultural Exchange and Knowledge Sharing: Preparing Humanity for Cosmic Citizenship

Another possible form of interaction is a gradual cultural exchange, in which extraterrestrials engage humanity in a controlled exchange of knowledge, technology, and values. This model assumes that humanity has reached a level of development sufficient to handle limited contact without societal collapse or ethical corruption. Under this scenario, extraterrestrials may openly introduce concepts and technologies that help humanity solve critical issues, such as energy shortages, disease, or environmental degradation, while carefully monitoring our use of these resources.

Controlled Technology Transfer for Global Advancement

If extraterrestrials choose to share technology with humanity, they might do so in a gradual, controlled manner to prevent misuse or dependency. The transfer of clean energy solutions, advanced medical knowledge, or environmental restoration techniques could be introduced selectively, allowing humanity to address pressing problems without compromising its autonomy. Extraterrestrials may view this as a stepping stone to future cooperation, providing humanity with the means to improve global living standards while testing our ability to use these advancements responsibly.

- **Comparison with Historical Alliances**: This approach resembles alliances between nations, where more advanced civilizations help developing societies through education, technology, and resources. Extraterrestrials may adopt a similar model, offering tools for sustainable development while ensuring humanity retains control over its own progress.

Cultural and Ethical Development: Learning Cosmic Values

Beyond technology, extraterrestrials might also seek to introduce new cultural and ethical concepts, guiding humanity toward a more universal morality. Messages that emphasize harmony with nature, respect for diverse life forms, and peace could be part of a gradual cultural exchange that elevates human values to align more closely with interstellar standards. This would involve a form of "cosmic citizenship training," in which humanity learns to see itself as one part of a larger, interconnected galactic society.

- **Example from Diplomacy and Cultural Exchange Programs**: Just as international exchange programs foster understanding and goodwill between nations, extraterrestrials might initiate cultural exchanges that expose humanity to cosmic perspectives on ethics, community, and the environment. These exchanges could gradually reshape human culture, creating a society more aligned with interstellar values.

Sudden Revelation and Open Contact: A Paradigm Shift

In a more radical scenario, extraterrestrials could decide to initiate open contact, revealing themselves in a manner that fundamentally changes human society. This "first contact" event could happen in various forms, ranging from public displays of extraterrestrial presence to mass communication through media or government channels. Such a paradigm shift would be transformative, requiring humanity to confront its beliefs, values, and place in the universe.

Public Contact and the Challenge of Assimilation

Open contact would bring immediate challenges as humanity grapples with the implications of extraterrestrial life and its advanced civilizations. Governments, religions, and social structures might face unprecedented questions, and humanity's sense of identity could undergo rapid transformation. Extraterrestrials might seek to ease this transition by providing information, offering guidance on social integration, or even establishing diplomatic relations to ensure that humanity navigates the upheaval in a constructive manner.

- **Historical Parallel with European Exploration**: The age of European exploration, which involved first contact between disparate cultures, often led to significant social and cultural changes for all parties involved. Open extraterrestrial contact could mirror this impact, but on a planetary scale, requiring careful management to prevent misunderstandings and conflicts.

Introducing Universal Laws and Cosmic Governance

In the wake of open contact, extraterrestrials might introduce universal laws or principles, encouraging humanity to align with ethical standards upheld by other civilizations. This could involve regulations related to environmental protection, respect for diverse life forms, and limitations on the use of dangerous technologies. As humanity transitions into a new era of cosmic awareness, extraterrestrials may view these guidelines as essential for maintaining stability and harmony within a broader interstellar community.

- **Example from the United Nations**: Just as the United Nations provides frameworks for international cooperation and peacekeeping, extraterrestrials might introduce frameworks for "cosmic governance." These principles would establish ethical norms that humanity would need to adopt in order to participate in interstellar affairs, balancing autonomy with shared responsibility.

A Path of Transformation: Humanity's Gradual Integration into an Interstellar Society

In a more optimistic scenario, extraterrestrials might view humanity as capable of transformative growth, choosing to help us integrate gradually into a larger galactic network. Rather than a sudden revelation or limited cultural exchange, this model envisions extraterrestrials working closely with humanity over time, assisting us in transforming our social structures, educational systems, and even our spiritual paradigms to align with cosmic values. This scenario suggests a period of mentorship in which extraterrestrials actively guide humanity through a transition from an isolated planet-bound society to a collaborative cosmic participant.

Creating a Shared Vision for the Future

In this model, extraterrestrials might inspire humanity to develop a shared vision of the future—one in which we overcome divisions, advance in science and spirituality, and collaborate on solving global challenges. They may provide resources, knowledge, or even "cosmic education" to help humanity reach a level of cultural maturity and ethical wisdom. By doing so, extraterrestrials could foster a vision for Earth that aligns with cosmic ideals, leading to a period of collective transformation and integration.

- **Comparison with Global Movements for Social Change**: Just as movements for human rights, environmentalism, and peace have transformed societies, extraterrestrials could catalyze similar changes on a planetary scale. Through their influence, humanity might achieve a new societal model rooted in cooperation, empathy, and a commitment to protecting both Earth and the cosmos.

A Harmonious Evolution: Humanity as a Cosmic Partner

In this transformative path, humanity would eventually become a full-fledged member of the galactic community, equipped to contribute its unique insights, creativity, and experiences. The goal of this mentorship would be to elevate humanity to a point where we are no longer recipients of guidance but active participants in a shared cosmic mission. Extraterrestrials who envision this future might focus on encouraging humanity's intrinsic strengths, nurturing a harmonious evolution in which our distinct cultural identity enriches the broader cosmic tapestry.

- **Analogy with International Collaboration**: Just as international collaboration enriches individual nations, humanity's integration into a cosmic society could bring benefits to both us and other civilizations. By contributing our unique perspectives and talents, humanity could add value to the collective knowledge, creativity, and resilience of the interstellar community.

Hypothetical Future Interactions and Humanity's Path Forward

Speculating on potential extraterrestrial interactions opens a window into the many possible futures humanity might face. From gradual, indirect guidance that fosters environmental and spiritual awareness to direct cultural exchanges that prepare humanity for interstellar citizenship, each scenario reveals a different vision of our potential place in the cosmos. Some scenarios involve sudden revelations and paradigm-shifting contact events that would require humanity to undergo rapid transformation, while others suggest a path of mentorship and gradual integration, allowing us to grow into our role as cosmic participants.

These hypothetical interactions raise profound questions about how humanity will respond to extraterrestrial influence. Will we be open to guidance, ready to embrace values of unity and peace? Or will we resist, clinging to outdated paradigms in the face of transformative opportunities? By examining these speculative futures, we gain not only insights into possible extraterrestrial intentions but also a deeper understanding of our own readiness for cosmic engagement and the role we may one day play within a larger interstellar society.

Interpretations and Theories About Extraterrestrial Intentions

Our exploration of extraterrestrial intentions offers a multifaceted understanding of why advanced civilizations might reach out to humanity. First, we considered the question of *Benevolent vs. Malevolent Intent,* analyzing extraterrestrial motivations and the ethical implications of interference. Some theories suggest a genuine concern for humanity's well-being, while others ponder potential self-interest or strategic aims. In examining these possibilities, we see that any guidance from extraterrestrials—if benevolent—may aim to prevent self-destruction, promote unity, or inspire us toward higher consciousness. On the other hand, more self-serving or even hostile intentions remind us to approach these unknowns cautiously.

Moving to the *Extraterrestrial Civilizations' Perspective on Humanity,* we explored hypotheses about how these beings might perceive our society. To some, we may be a younger civilization worthy of mentorship or study, while others may view us with caution, given our tendency toward conflict and environmental harm. The exploration of why extraterrestrials might want to intervene reveals that they may see in us both potential allies and risks, raising questions about what they might hope to achieve by intervening or providing guidance. Their actions could be rooted in a broader commitment to cosmic harmony, or, alternatively, a sense of moral obligation to protect life.

Finally, our discussion of *Speculative Theories on ETs and Earth's Future* led us to consider hypothetical roles extraterrestrials might play in shaping our evolution. These include potential biological or technological influence, mentorship to prepare us for interstellar citizenship, or indirect guidance fostering spiritual and ethical growth. We also examined possible future scenarios of interaction, from controlled cultural exchanges to transformative contact events. These scenarios illustrate how extraterrestrials may see their guidance as part of a larger cosmic plan, aiding humanity's transition into a broader interstellar society or preserving Earth as a unique civilization with valuable perspectives.

In sum, these topics provide a rich blend of speculative and philosophical insights into why and how extraterrestrials might interact with humanity. Whether as mentors, observers, or potential partners, extraterrestrials' intentions and methods reveal much about both their values and our own. As we consider these possibilities, we are invited to reflect on humanity's readiness for cosmic engagement, our ethical responsibilities, and the paths we might choose as we encounter beings from beyond our world.

As we delve deeper into understanding extraterrestrial messages and their potential significance, it's essential to address a fundamental question: how credible are these encounters, and what can we trust in the reports? Analyzing the *Credibility and Authenticity of Messages* provides a critical foundation for distinguishing genuine experiences from those influenced by psychological, social, or even governmental factors. By examining witness credibility, we assess the backgrounds, testimonies, and possible corroborations among individuals who report extraterrestrial encounters, as well as the psychological influences that might impact their claims. Beyond personal accounts, we will explore the physical evidence surrounding these events, including soil samples, radar data, and the reliability of photographic and video documentation, which can add or detract from a case's legitimacy. Finally, we'll investigate the role of governments, examining declassified documents and the persistent theories of cover-ups and secrecy, which suggest that certain knowledge about extraterrestrial phenomena may be intentionally concealed. This assessment will guide us in building a more informed perspective on the authenticity of extraterrestrial messages and the broader implications of these encounters.

Chapter 8: Unveiling the Truth: Evaluating the Credibility and Authenticity of Extraterrestrial Messages

When it comes to claims of extraterrestrial messages, determining their authenticity is critical to understanding the phenomenon itself. Are these encounters based on genuine experiences, or are they influenced by psychological, social, or even governmental factors? In this section, we explore the credibility of reported messages by examining key elements: the reliability of witnesses, the solidity of physical evidence, and the potential for government involvement or suppression of information. We begin with an in-depth look at witness credibility, considering both the backgrounds of individuals who report encounters and the psychological or social dynamics that could affect their testimonies. Next, we assess physical evidence like radar data, soil samples, and video footage, discussing both the strengths and limitations of these forms of documentation. Finally, we turn our attention to the role of governments, evaluating declassified documents and theories suggesting secrecy and cover-ups in order to understand the possible agendas behind them. By carefully analyzing each of these facets, we can develop a more nuanced perspective on the reliability of extraterrestrial messages and the forces that may shape public perception of these phenomena.

Examining the Source: Assessing Witness Credibility in Extraterrestrial Encounters

When evaluating the authenticity of extraterrestrial messages, assessing witness credibility is a critical first step. A witness's background, the consistency of their story, and corroboration from multiple sources can provide insight into the reliability of reported encounters. But credibility assessment is a nuanced process—one that goes beyond simply verifying an individual's reputation or the coherence of their testimony. In the field of UFO studies, witness credibility has been examined in cases involving individuals from diverse backgrounds, including professionals in high-stakes fields like aviation and the military, where observational accuracy is crucial. Understanding how these elements contribute to the reliability of extraterrestrial claims helps us form a more solid foundation for evaluating the larger phenomenon.

The Importance of Background: Does a Witness's Profession Matter?

One of the primary elements in evaluating witness credibility is their background, including occupation, social status, and education. Witnesses with backgrounds in fields that prioritize observational skills—such as pilots, scientists, and law enforcement—are often perceived as more reliable due to their trained attention to detail and adherence to factual reporting. For instance, when a commercial airline pilot reports an unidentified aerial phenomenon (UAP), their experience with the skies lends weight to their testimony. Similarly, military personnel are often seen as credible witnesses because of their familiarity with advanced technologies and aircraft.

High-Stakes Observers: Aviation and Military Personnel

Pilots and military officials represent some of the most compelling witnesses in UFO cases, as their professional experience is inherently linked to high-stakes observation and accuracy. Cases like the 1986 Japan Airlines Flight 1628 encounter, in which a seasoned pilot reported a massive UAP following his aircraft, highlight how such witnesses can lend credibility to a sighting. Their ability to accurately gauge distances, altitudes, and flight behavior is crucial in providing detailed accounts. Additionally, the military's standardized procedures for reporting unknown objects allow for detailed and often corroborated accounts that are useful for further investigation.

- **Case Study**: In 2004, the U.S. Navy's Tic Tac UFO incident involved multiple trained personnel on the USS Princeton and the USS Nimitz who corroborated an encounter with a fast-moving, highly maneuverable UAP. The presence of skilled observers in both air and naval forces contributed to the case's credibility, leading to further investigation by the U.S. Department of Defense.

Everyday Observers: The Role of Civilian Witnesses

While professionals are often seen as more credible, civilian witnesses make up a significant portion of extraterrestrial reports. Here, credibility is often judged on the witness's ability to provide a coherent, consistent, and unemotional account of their experience. Civilian sightings, like the famous 1961 Betty and Barney Hill abduction case, have generated much interest due to the detailed nature of the accounts, despite the witnesses lacking formal observational training. However, skepticism toward civilian accounts often focuses on the potential for imagination, social influence, and psychological factors to play a larger role.

Corroboration: Strength in Numbers

Credibility increases significantly when multiple witnesses provide similar accounts of an encounter. Corroborated stories, particularly from unrelated individuals, add weight to the authenticity of a reported message or event. Witness corroboration has been a key factor in high-profile UFO cases, where various people independently describe the same event. When different people report similar observations—whether in physical descriptions of extraterrestrials, details about spacecraft, or the content of telepathic messages—the reliability of the case strengthens.

Public Sightings and Mass Witness Events

Cases involving large numbers of witnesses are particularly compelling, as they reduce the likelihood that the event was imagined or misinterpreted. The 1997 Phoenix Lights incident, for example, was observed by hundreds of people across Arizona, including police officers and public officials, who described similar details of a massive V-shaped craft. Such widespread corroboration presents a strong case for authenticity, especially when witnesses are located in different areas and provide independent accounts that match.

- **Case Study**: The 1994 Ariel School encounter in Zimbabwe involved 62 schoolchildren who claimed to see a UFO and received telepathic messages from extraterrestrial beings. Each child provided a strikingly similar description, lending credibility to the account despite the young age of the witnesses. This incident demonstrates the power of multiple corroborative testimonies to support the credibility of extraordinary events.

Independent Accounts: The Importance of Isolation in Witness Corroboration

One of the strongest indicators of credibility is when independent witnesses, unknown to each other, describe the same or similar events. This phenomenon suggests that the event was not influenced by social contagion or group imagination. In the Rendlesham Forest incident of 1980, U.S. Air Force personnel stationed at RAF Woodbridge in England provided independent accounts of strange lights and physical evidence that indicated a landing. Each witness reported consistent details despite being isolated from others during the experience, enhancing the case's credibility and prompting further investigation.

Consistency Over Time: Examining Testimonial Stability

The consistency of a witness's account over time is a crucial aspect of credibility. A credible witness is likely to provide a steady narrative without significant changes, even when questioned under stress or after prolonged intervals. In some cases, witnesses have maintained consistent accounts for decades, reinforcing the authenticity of their experiences.

Stability Through Repeated Interviews

Witnesses who undergo multiple interviews—particularly with different interviewers—are often scrutinized for consistency. Discrepancies or changes in their stories can weaken the credibility of their accounts. However, if a witness's narrative remains consistent over multiple retellings, their reliability as a source is strengthened. For example, Betty Hill's account of her abduction experience was repeatedly investigated, including under hypnosis, and her story exhibited strong consistency, which has contributed to its place as a landmark case in ufology.

Psychological Stress and Memory Distortion

While consistency is a key indicator of credibility, it's also important to consider the psychological impact of traumatic or extraordinary experiences on memory. Witnesses may experience shifts in perception or recall under stress, especially when their accounts involve frightening or life-changing encounters. Such factors need to be evaluated with sensitivity, as slight inconsistencies may not necessarily discredit the witness but instead reflect the mental impact of their experience.

- **Psychological Perspective**: Researchers suggest that memory may sometimes alter details subconsciously in response to trauma. Rather than indicating deception, minor changes in accounts may reveal the brain's attempt to process extraordinary events. For credible witnesses, these variations are typically minor and do not affect the overall coherence of the account.

Verifying Backgrounds: Investigative Techniques in Credibility Assessment

An essential part of assessing witness credibility involves verifying their backgrounds, including employment history, educational background, and psychological health. Investigators often look for information that may explain a motive for deception or suggest a predisposition toward imaginative or attention-seeking behavior. This step helps discern whether a witness is more likely to report an encounter accurately or to fabricate or exaggerate details.

Employment and Educational Records

In cases involving high-profile witnesses, such as pilots or government officials, investigators often verify employment records to confirm the witness's role and any relevant expertise. This process also serves to ensure that the witness has a solid understanding of their field, lending weight to their observational abilities and reducing the likelihood of misinterpretation. Educational background can also provide insight, as individuals with scientific or analytical training may be more likely to provide a reliable account.

Psychological Assessment: Understanding Witness Mindset

In some instances, investigators may look into a witness's psychological history, particularly if they exhibit signs of trauma or stress related to the encounter. While this is a delicate area, understanding a witness's psychological profile can reveal potential biases or conditions that might influence their perceptions. However, it's essential to approach this factor with sensitivity, as mental health status does not automatically discredit a witness. Some witnesses report experiences that fit within normal psychological bounds and show no signs of predisposition toward imaginative beliefs or exaggeration.

Building a Reliable Picture Through Witness Credibility

Assessing the credibility of witnesses is a foundational step in evaluating extraterrestrial messages and encounters. By examining witness backgrounds, corroborative accounts, consistency over time, and investigative verification, we build a comprehensive understanding of what contributes to a reliable testimony. High-stakes professionals like pilots and military personnel often bring added weight to their accounts, given their trained observational skills. Meanwhile, corroborative cases involving multiple independent witnesses or mass sightings lend collective strength to claims, reducing the possibility of misinterpretation or fabrication.

As we continue to explore extraterrestrial messages, understanding witness credibility allows us to sift through a complex web of personal accounts with greater discernment. By combining thorough background checks, corroboration, and an appreciation for the psychological impact of extraordinary events, we can construct a clearer, more credible picture of these encounters, guiding us in our search for authentic, potentially transformative messages from beyond Earth.

Understanding the Mind: Psychological and Social Factors Affecting Witness Credibility in Extraterrestrial Encounters

The human mind is complex, shaped by psychological patterns, social influences, and cultural contexts that can all impact how we perceive and recount extraordinary experiences. When it comes to evaluating witness credibility in extraterrestrial encounters, examining psychological and social factors provides valuable insights into how individuals might interpret, remember, and communicate these events. By understanding these factors, we gain a fuller picture of the strengths and potential pitfalls in witness testimonies, moving beyond a simple acceptance or rejection of their stories to a more nuanced appreciation of the human experience in confronting the unknown.

The Power of the Mind: Psychological Influences on Credibility

A witness's psychological state can profoundly influence their perception and interpretation of an encounter with the unexplained. Psychological responses to stress, memory processes, and individual personality traits all play a role in how someone might experience and later recount an encounter. These factors can both strengthen and undermine the credibility of a testimony, depending on how they manifest in the witness's account.

Memory, Perception, and the Impact of Extraordinary Events

Memory is not a static, perfectly accurate record of past events. Instead, it is a dynamic process, prone to distortions, especially when dealing with high-stress or emotionally charged experiences. In the case of extraterrestrial encounters, witnesses often report feeling shock, fear, or awe, emotions that can affect how details are encoded and later recalled. Psychologists have found that during intense experiences, memory may be enhanced for central details but may lose clarity around peripheral aspects, which can lead to an account that feels fragmented or inconsistent.

- **Example**: In cases of abduction experiences, witnesses often report vivid, almost photographic recollections of specific elements, such as the appearance of beings or the sensation of paralysis, while details surrounding the event's start or end may appear vague. This phenomenon, sometimes called "flashbulb memory," suggests that intense experiences might create strong but selective memories that influence credibility without implying deceit.

The Role of Suggestibility and Hypnotic Recall

Some witnesses undergo hypnosis to recall details of their encounters, especially in cases where memories are fragmented or repressed. However, hypnosis itself can influence memory through suggestion, a phenomenon where the hypnotist's questions or the witness's expectations shape the details of the recall. Suggestibility can introduce subtle changes or add imagined details to a memory, which complicates the task of distinguishing between a genuine recollection and elements that might have been unintentionally planted.

- **Case Study**: The famous Betty and Barney Hill abduction case involved hypnotic recall, where each partner described being taken aboard a spacecraft. While the consistency between their stories lends credibility, skeptics have argued that the hypnotist's questioning could have influenced their narratives. Hypnosis, therefore, is a double-edged sword: it can retrieve memories that might otherwise be lost but can also make witnesses more prone to suggestion.

Personality Traits and Belief Systems: How They Shape Perception

An individual's personality traits and belief systems can shape how they interpret an encounter and even influence their likelihood of perceiving one. People with a strong openness to new experiences may be more inclined to see unusual lights or shapes in the sky as extraterrestrial, while those with high levels of skepticism may dismiss similar experiences as mundane phenomena. Psychological traits like openness, neuroticism, and suggestibility can thus contribute to how credible a witness's testimony might appear to an investigator.

- **Example**: A witness who holds strong paranormal beliefs may be more likely to interpret ambiguous stimuli, such as a distant light, as a spacecraft, while a skeptic might interpret the same light as an aircraft. Investigators must consider these individual differences, as they can lead to honest yet subjective interpretations that may diverge significantly from the actual events.

Social Influences and the Shaping of Testimony

Beyond individual psychology, social factors also play a powerful role in shaping witness credibility. The influence of group dynamics, cultural beliefs, and societal expectations can shape how a witness perceives and reports an encounter. These influences can affect not only what a person remembers but also how they choose to communicate it, as social acceptance or skepticism can create pressure to align with prevailing attitudes.

Group Encounters and the Influence of Collective Belief

When multiple people witness an extraterrestrial event together, a phenomenon known as "group reinforcement" can take place, where individuals align their perceptions with the dominant narrative of the group. While this can strengthen credibility by providing corroboration, it can also introduce bias as individuals subconsciously conform to each other's accounts. Group encounters, like mass sightings, often yield more credible testimonies, yet they must be carefully examined for signs of social reinforcement.

- **Case Study**: In the Ariel School encounter in Zimbabwe, schoolchildren who witnessed a UFO described receiving telepathic messages. Their testimonies were remarkably consistent, but the possibility of social reinforcement remains a factor, as the children may have been influenced by each other's descriptions. However, the strong consistency across independent accounts also suggests authenticity, highlighting the need for a balanced approach in assessing group encounters.

Cultural Beliefs and Their Impact on Interpretation

Cultural context plays a vital role in how people interpret encounters with the unknown. For instance, societies with strong spiritual or animistic traditions may view extraterrestrial beings as spiritual entities or gods, while cultures with a scientific orientation might interpret them as advanced alien civilizations. Cultural influences can shape not only the interpretation of the event but also the vocabulary and symbols used to describe it, which affects how the story is received by others.

- **Example**: In Native American cultures, stories of "star people" visiting Earth have existed for generations, often interpreted through a spiritual or mythological lens. Such cultural framing can lend credibility to witnesses within those societies but may introduce interpretative biases when assessed by those from different cultural backgrounds. Investigators must therefore consider the cultural lens through which a witness's experience is filtered, recognizing that cultural context can add both richness and complexity to the account.

Fear of Judgment and the Pressure to Conform

Witnesses of extraterrestrial encounters often face skepticism or ridicule, which can influence how openly they share their experiences. Fear of social judgment may lead some individuals to withhold or alter details, affecting the perceived credibility of their testimonies. In certain cases, witnesses may even withdraw their claims or adjust their stories to make them more acceptable to a skeptical audience. Conversely, some may exaggerate details to gain social validation or attention.

- **Case Study**: Military personnel involved in the 1980 Rendlesham Forest incident faced significant skepticism and institutional pressure, which may have influenced how they reported their experiences. While some persisted in their accounts despite potential backlash, others became reluctant to discuss the event openly. This highlights how social factors can pressure witnesses to alter or downplay their testimonies, ultimately impacting the account's credibility.

Social Media and the Spread of Encounter Narratives

In today's digital age, social media has a powerful impact on witness credibility, enabling witnesses to share their stories widely and receive validation or skepticism from a global audience. Social media can amplify certain narratives, often leading to a rapid spread of encounter reports. While this can help witnesses find support, it also increases the risk of embellishment or conformity as individuals seek social validation in an environment where exaggeration is often rewarded.

- **Example**: Recent mass sightings, like the Chicago Mothman reports in 2017, spread quickly on social media, where witnesses and enthusiasts shared photos, descriptions, and commentary. While this immediate dissemination can bolster credibility through corroboration, it can also create an echo chamber effect, where repeated retellings modify or exaggerate details. Investigators must remain cautious in considering such accounts, recognizing that social media can shape both perception and credibility in powerful ways.

The Interplay of Psychological and Social Factors in Assessing Credibility

Understanding witness credibility in extraterrestrial encounters requires an appreciation of the complex psychological and social influences that shape perception and memory. Psychological factors such as memory processes, suggestibility, and personality traits can impact how encounters are experienced and recalled, creating both strengths and potential weaknesses in testimonies. Social influences, from group dynamics to cultural beliefs and social media, further affect how witnesses interpret and communicate their experiences, sometimes enhancing credibility and other times introducing bias.

By examining these psychological and social elements, investigators can approach witness testimonies with a more nuanced perspective, distinguishing between genuine experiences and those shaped by cultural, social, or mental factors. This balanced approach allows for a deeper understanding of the human dimension in extraterrestrial encounters, bridging the gap between skepticism and belief while honoring the rich complexity of individual experiences with the unknown.

Tangible Traces: Evaluating Physical Evidence in Extraterrestrial Encounters

While witness accounts provide valuable insights, physical evidence adds a level of credibility that can substantiate the extraordinary claims associated with extraterrestrial encounters. Unlike memories or subjective experiences, physical evidence offers material that can be scientifically analyzed, tested, and potentially validated. This category of evidence includes items like soil samples from landing sites, radar data tracking unidentified objects, and even unusual biological samples, each adding a layer of credibility to encounters. Examining these forms of evidence allows us to move beyond human perception into the realm of measurable data, building a more solid foundation for evaluating the authenticity of extraterrestrial messages.

Grounded Clues: Soil Samples from Alleged Landing Sites

Soil samples collected from alleged UFO landing sites can reveal chemical and physical anomalies that point to extraordinary events. Investigators look for unusual levels of radiation, altered soil composition, and changes in magnetic properties that may indicate exposure to an unknown energy source or material. These traces provide a concrete starting point for understanding extraterrestrial presence, offering data that can be compared against control samples to determine whether an external, non-natural force was involved.

The Science Behind Soil Analysis in UFO Investigations

Soil analysis is a multi-layered process involving a variety of scientific techniques. Geochemical tests, for example, can reveal unusual mineral concentrations, while spectroscopic analysis identifies chemical changes at the atomic level. Thermoluminescence tests can detect whether the soil has been exposed to intense heat. Such findings can indicate the presence of an intense energy source—perhaps a spacecraft engine or other technology beyond known human capability. When results show significant deviations from normal soil composition, it suggests the interaction of the environment with an unknown force, thereby lending credibility to witness claims of a landing event.

- **Case Study: Delphos, Kansas, 1971**: A well-documented case involved a circular patch of soil that remained dry and barren for years following a UFO sighting. Analysis revealed chemical changes in the soil structure and unusually high levels of phosphates and oxalates, which investigators suggested could have been caused by a high-temperature event. The site remained resistant to water absorption and plant growth, reinforcing the possibility of an extraordinary incident.

Assessing Radiation and Magnetic Changes

One of the most compelling pieces of evidence in soil samples is the presence of radiation. Elevated radiation levels, measured against background radiation in nearby areas, can suggest exposure to radioactive material or an energy source not typically found in nature. In some cases, soil samples have shown traces of isotopes rarely found in the Earth's crust, leading scientists to consider the involvement of unknown materials or technology. Additionally, magnetic anomalies, often revealed through magnetometer readings, may indicate exposure to a powerful electromagnetic field, another possible signature of advanced technology.

Tracking the Skies: Radar Data as an Objective Record

Radar data provides an objective, real-time record of unidentified aerial phenomena, allowing investigators to track the speed, altitude, and trajectory of unknown objects. Unlike eyewitness accounts, radar data is a direct measure of an object's physical properties, providing information that is difficult to dispute. Recorded radar signatures from UFO encounters often reveal flight characteristics that defy known technology, such as rapid acceleration, sharp turns, and abrupt stops that would be impossible for conventional aircraft.

Anomalies in Speed, Altitude, and Trajectory

Radar data has repeatedly recorded objects moving at speeds beyond any known aircraft, performing maneuvers that would generate forces far beyond human tolerance. In many instances, radar operators have noted objects traveling at thousands of miles per hour, descending from high altitudes to near ground level almost instantly. Such capabilities suggest technologies that far exceed current human engineering and reinforce the idea of extraterrestrial intelligence behind these encounters.

- **Case Study: The 2004 Nimitz Encounter**: In this well-documented incident, radar systems on the USS

Princeton tracked a fast-moving object later known as the "Tic Tac" UFO. The object displayed characteristics beyond known technology, including sudden shifts in altitude from 80,000 feet to sea level in less than a second. Such movements are impossible under the laws of physics as we understand them, making radar data from this case compelling evidence for further investigation.

Corroborating Radar Data with Visual and Infrared Sensors

Radar data gains additional credibility when supported by other sensory data, such as visual sightings and infrared (IR) signatures. Infrared tracking provides insight into heat emissions from the object, which can be analyzed to estimate the energy output or propulsion mechanisms. When visual sightings, infrared sensors, and radar signatures all indicate the presence of an unknown object, the case gains considerable strength, as the combination of independent data points reduces the likelihood of equipment error or natural phenomena.

- **Example**: In several military encounters, such as those involving F/A-18 pilots, radar data has been corroborated by IR sensors detecting heat anomalies around the objects. This data suggests propulsion methods that produce intense energy yet do not conform to any known propulsion technology. Multiple forms of sensory input create a fuller picture, adding layers of verification that are difficult to dismiss.

Biological and Material Residues: Evidence Beyond the Soil and Skies

In some encounters, physical evidence includes biological or other material residues believed to be left by extraterrestrial beings or their technology. Analysis of such materials can be particularly illuminating, as they may contain chemical or biological signatures that differ from terrestrial norms. Material residues often consist of unusual metal alloys, synthetic substances, or even alleged biological samples that can undergo laboratory analysis.

The Search for Extraterrestrial Alloys and Synthetic Materials

Extraterrestrial materials often include alloys or synthetics with properties that challenge known metallurgical techniques. Some samples, upon examination, show elements combined in ways that are not typically possible with earthly technology, such as isotopic ratios that differ significantly from those found in Earth's crust. These materials can be subjected to spectroscopic analysis, electron microscopy, and other advanced techniques to reveal details about their composition, offering hints at possible extraterrestrial origins.

- **Case Study: The Ubatuba Magnesium Fragments**: In 1957, fragments believed to be from a UFO were recovered from a beach in Ubatuba, Brazil. These fragments were composed of nearly pure magnesium, with isotopic ratios that suggested an origin outside of Earth. Although the origins remain disputed, the samples provide a compelling avenue for investigation, as pure magnesium in that concentration and form is difficult to produce with natural terrestrial processes.

Biological Evidence: DNA and Cellular Analysis

In rare cases, alleged biological samples have been recovered and analyzed for cellular structure, genetic markers, or unique chemical compositions. While such evidence is rare and often contentious, it offers a potentially groundbreaking form of proof. Any biological material with DNA markers distinctly different from earthly organisms would have profound implications for understanding extraterrestrial life. However, due to contamination risks and the controversial nature of such findings, this evidence requires rigorous protocols and careful examination.

- **Example**: Some abduction cases have included reports of biological samples, such as skin or hair, allegedly left by extraterrestrial beings. These samples have occasionally shown anomalies in cellular structure or DNA that raise questions about their origin. While not universally accepted as definitive evidence, these biological traces, if analyzed thoroughly, could potentially support claims of extraterrestrial encounters.

Interpreting Anomalous Evidence: Limitations and Challenges

While physical evidence offers valuable insight, interpreting it comes with challenges. Contamination, environmental factors, and equipment limitations can complicate analysis. Soil samples may be affected by weather or previous human activity, and radar data can sometimes be influenced by atmospheric conditions. To ensure reliability, investigators must carefully account for these variables, often by comparing results from multiple samples or repeated radar readings.

Contamination and Environmental Influence

Contamination is a major concern, particularly with soil or biological samples that could be influenced by environmental factors, such as nearby human activity or natural events. Laboratories specializing in extraterrestrial evidence take great precautions to prevent contamination by using sterilized tools, controlled environments, and rigorous sample handling procedures. By comparing the anomalous sample with nearby control samples, investigators can better determine whether the observed changes are unique to the site of the encounter or are part of natural variations in the environment.

Equipment and Measurement Limitations

In the case of radar and infrared data, measurement limitations can sometimes create false positives. Atmospheric interference, for example, may occasionally create radar signatures that resemble unidentified objects. However, these limitations are generally understood by trained operators, who use established protocols to rule out false readings. Multiple, independent radar readings further help to confirm the presence of anomalous objects, enhancing the reliability of the data.

Building a Case Through Physical Evidence

Evaluating physical evidence is essential for building credibility in extraterrestrial encounters. Soil samples, radar data, and material residues each provide tangible clues that can substantiate witness claims, creating a multi-layered approach to assessing these phenomena. Soil samples with unusual radiation levels or altered chemical compositions point to an extraordinary energy source, while radar data offers objective tracking of anomalous objects displaying movements beyond human technology. Material and biological residues add an additional layer of mystery, often containing unique compositions that suggest non-terrestrial origins.

While challenges remain, such as contamination risks and measurement limitations, careful analysis and corroborative evidence help strengthen the case for authenticity. By grounding witness testimonies in tangible, scientifically measurable evidence, we gain a more comprehensive view of extraterrestrial encounters—one that bridges subjective experience with objective investigation and paves the way for a deeper understanding of the unknown.

Capturing the Unknown: Evaluating the Reliability and Limitations of Photographic and Video Evidence

Photographic and video evidence has long been a key focus in the study of extraterrestrial encounters. Unlike anecdotal testimony, these forms of evidence provide a visual record that can be analyzed repeatedly, allowing for technical scrutiny and closer inspection of alleged sightings or encounters. However, the reliability of photographic and video evidence is affected by multiple factors, such as technology limitations, environmental conditions, and the potential for manipulation. While these visual records can powerfully support witness accounts, they also pose unique challenges in authenticity assessment.

The Evolution of Visual Evidence: A Brief History of UFO Photography and Videography

Since the advent of cameras, individuals have captured images they believe depict extraterrestrial phenomena. Early UFO photos, taken as far back as the 1940s, were limited by the film technology of the time, resulting in grainy, low-resolution images that are difficult to authenticate. The development of digital photography and high-definition video has enhanced the quality of visual evidence, but it has also introduced new complications, such as digital manipulation and interpretive biases introduced by advanced editing tools. Evaluating the credibility of visual evidence requires understanding both the history of these technologies and the ways in which they can mislead or support our interpretations.

- **Case Study**: The McMinnville UFO photographs from 1950, taken by a farmer in Oregon, remain some of the most iconic UFO photos. Although highly scrutinized, these black-and-white images continue to intrigue researchers due to their clarity and the simplicity of the technology used, which minimized the risk of manipulation.

Assessing the Reliability of Photographic Evidence

In order to determine the reliability of a photograph as evidence, investigators examine several technical aspects, including resolution, image quality, environmental conditions, and the authenticity of the medium. High-resolution images allow for closer inspection of details such as shape, texture, and surrounding objects, whereas low-resolution images often create interpretive ambiguity. Additionally, environmental factors, such as lighting conditions and potential reflections, can influence how an object appears on film, leading to misinterpretations or misidentifications.

The Importance of High Resolution and Clarity

Higher resolution enhances the quality of analysis by allowing investigators to zoom in on specific features without losing clarity. Detailed images help reveal characteristics that distinguish unknown objects from familiar aircraft or natural phenomena, such as cloud formations. In cases where UFOs exhibit unusual shapes or reflective surfaces, high resolution can provide clues about the material composition or design, helping to rule out conventional explanations.

- **Example**: The 2007 drone UFO photos from Capitola, California, featured highly detailed images of an unusual craft with intricate designs. The high resolution led to extensive analysis by enthusiasts and skeptics alike, although the photos were later debated for potential digital manipulation. This case highlights how high resolution, while beneficial, does not guarantee authenticity but provides more data for analysis.

Environmental Conditions: Light, Weather, and Perspective

Lighting conditions and atmospheric factors play a significant role in how objects appear in photographs. Bright sunlight, lens flare, and glare can distort the appearance of an object, while fog or mist may create optical illusions. When analyzing photographic evidence, investigators must consider these conditions to determine whether they may have affected the image's appearance, as certain lighting effects can create shapes or reflections that resemble UFOs.

- **Example**: The Phoenix Lights event of 1997, witnessed by thousands, was captured in both photos and videos. Some images appeared to show a V-shaped craft, but atmospheric haze and city lights created challenges in interpreting the true nature of the lights. In this case, visual records needed to be cross-referenced with witness accounts and other evidence to build a coherent understanding of the event.

Video Evidence: Movement and Behavior Analysis

Video footage offers an additional layer of information by capturing the movement and behavior of unidentified objects. Unlike still images, videos can show speed, trajectory, and changes in direction, providing clues about the capabilities and potential origin of a craft. Analyzing video evidence requires examining frame-by-frame details and noting any unusual movements that defy the capabilities of known technology, such as abrupt stops, rapid accelerations, or unexplainable changes in altitude.

Tracking Speed, Acceleration, and Maneuverability

Unusual motion patterns are often considered a hallmark of credible UFO sightings, especially when the object exhibits behaviors that would generate extreme G-forces beyond human tolerance. Advanced software can measure the speed and trajectory of objects within video footage, giving analysts concrete data on the object's performance. When such behavior exceeds known human technology, it strengthens the case for an unknown, possibly extraterrestrial origin.

- **Case Study**: The 2004 Nimitz "Tic Tac" encounter, captured on video by U.S. Navy pilots, showed an object moving at rapid speeds and making sharp turns. The video's release led to significant analysis, as the maneuvers appeared to exceed the capabilities of any known aircraft, making it one of the most credible pieces of video evidence in recent years.

The Role of Frame Rate and Image Stabilization

Frame rate and stabilization are essential for ensuring video clarity, particularly when the object is moving quickly or when the footage is taken from a moving platform, such as an aircraft or a moving vehicle. Low frame rates can create choppy footage, making it difficult to track the object smoothly. Stabilization tools can help create a clearer view but may introduce distortion if overused. These technical considerations influence the quality of analysis and the conclusions that can be drawn about the object's movement and behavior.

Limitations of Photographic and Video Evidence

Despite their potential to support claims of extraterrestrial encounters, photographic and video evidence comes with limitations. Digital manipulation, environmental misinterpretations, and equipment limitations pose challenges to verifying authenticity. Understanding these limitations is crucial for balanced evaluation, as even the most compelling visual evidence must be scrutinized for signs of tampering or environmental influence.

The Challenge of Digital Manipulation

In the digital age, photo and video editing software makes it possible to alter or create UFO images and footage with remarkable realism. Techniques like compositing, digital filtering, and artificial lighting effects allow for convincing hoaxes. Authenticating digital evidence requires forensic analysis to detect signs of manipulation, such as inconsistencies in lighting, shadows, or pixel structure. Analysts often look for "noise patterns" within the image, as natural and digitally manipulated sections of a photo will typically show different textures.

- **Example**: In 2011, a supposed UFO video from Jerusalem showed a bright orb hovering over the city's Dome of the Rock. While the video initially garnered attention, analysts later determined that the video had likely been digitally altered, as different versions showed inconsistencies in movement and lighting, indicating a probable hoax.

Environmental Misinterpretations and Optical Illusions

Environmental factors, such as lens flares, reflections, and light distortions, can produce visual effects that resemble UFOs. Additionally, certain camera angles and optical illusions, like parallax, can distort the perceived distance and size of objects, leading to misidentification. By comparing the image with known objects or landmarks in the background, investigators can often determine whether the phenomenon is the result of natural factors rather than an unknown object.

- **Example**: Many alleged UFO sightings turn out to be reflections or flares caused by the sun hitting the camera lens at a particular angle. Some famous photos, like those from the Hessdalen Lights in Norway, have been attributed to unique atmospheric conditions that produce recurring light phenomena, which are often mistaken for UFOs.

Equipment Limitations and Artifacts

Older cameras, as well as certain digital models, can produce artifacts—small distortions or blurs—especially when capturing moving objects in low light. Artifacts can lead to mistaken interpretations of shapes and sizes, creating false impressions of extraterrestrial objects. Advanced analysis techniques, such as image stacking and digital noise reduction, can help to isolate the true image from equipment-related distortions.

Advances in Forensic Analysis of Visual Evidence

Forensic analysis has evolved to provide more sophisticated methods for authenticating visual evidence, especially digital images and videos. Analysts use metadata analysis, shadow mapping, and 3D modeling to confirm whether an object's size, shape, and lighting match its environment. Metadata embedded within digital files, such as the time and location of capture, can reveal discrepancies that indicate tampering. Techniques such as these allow for a more rigorous, science-based approach to evaluating visual evidence.

Metadata Analysis for Authenticity Checks

Metadata contains crucial information about an image or video, including details about the camera used, the timestamp, and sometimes even the GPS location. By examining metadata, investigators can confirm whether an image was taken at the reported location and time or if it has been altered. However, metadata can also be edited, so it must be cross-checked with other forms of evidence.

Shadow Mapping and 3D Modeling

Shadow mapping involves analyzing the direction, length, and consistency of shadows within an image to determine if they align with the position of light sources at the reported time and location. Similarly, 3D modeling techniques allow investigators to recreate the environment in a digital space, comparing the alleged UFO's dimensions and positioning with known surroundings. These methods help rule out discrepancies that might indicate a staged or edited image.

Balancing Promise and Pitfalls in Visual Evidence

Photographic and video evidence offers a powerful way to document and analyze alleged extraterrestrial encounters. High-resolution images and video footage capturing an object's speed, trajectory, and movement characteristics provide tangible data that can substantiate witness claims. However, limitations in technology, environmental conditions, and the potential for digital manipulation require a careful, systematic approach to analysis. Advanced forensic techniques, such as metadata examination and shadow mapping, help investigators assess authenticity, but even these tools must be applied with caution.

Ultimately, photographic and video evidence plays a valuable role in the evaluation of extraterrestrial encounters, offering visual proof that complements other forms of evidence. By understanding both the potential and limitations of this evidence, we gain a more balanced view of what images and videos can reveal about the mysteries of the unknown and how they contribute to the broader understanding of extraterrestrial phenomena.

The Hidden Files: Analyzing Declassified Documents and Government Involvement in Extraterrestrial Encounters

Government involvement in the investigation of UFOs and extraterrestrial encounters has long fascinated researchers, raising questions about what information might be hidden from public view. Declassified documents offer a unique window into the world of official UFO investigations, revealing the scope and focus of governmental interest in the phenomenon. These documents can provide insight into cases that were thoroughly investigated, the methods used, and the level of concern expressed by various branches of the military and intelligence agencies. While they sometimes affirm public suspicions of governmental secrecy, they also bring clarity to the genuine interest and, occasionally, the caution with which officials approached these unexplained events.

The Beginnings of Disclosure: Project Blue Book and Its Legacy

The United States government's first significant effort to systematically investigate UFOs began with Project Blue Book, initiated by the U.S. Air Force in 1952 and continuing until 1969. This program aimed to study and catalog UFO sightings, ultimately aiming to determine whether they posed a threat to national security. Project Blue Book generated thousands of pages of reports and analyses, many of which have since been declassified. Although the project officially concluded that most sightings could be attributed to natural phenomena or human error, a small percentage of cases remained unexplained, fueling public speculation about the government's true knowledge.

Examining Project Blue Book's Findings and Limitations

Declassified Blue Book files include reports on thousands of sightings, accompanied by investigative notes, photographs, and, in some cases, witness interviews. Many sightings were dismissed as weather balloons, aircraft, or astronomical objects. However, the few cases that could not be explained became the cornerstone of UFO research, suggesting the existence of a phenomenon that defied easy categorization. For example, sightings by trained military personnel were often marked as more credible, given their expertise in identifying aircraft and atmospheric phenomena.

- **Case Study: The 1952 Washington D.C. Sightings**: Declassified files from Blue Book include reports of radar and visual sightings over Washington, D.C., which triggered national media coverage and sparked a formal investigation. While the Air Force attributed these sightings to "temperature inversions," many in the public and the UFO research community found this explanation unconvincing. This case remains one of the most notable entries in the Blue Book archives and is often cited as an example of government reluctance to fully disclose the nature of some encounters.

Project Blue Book's Closure and the Transition to Secrecy

After Project Blue Book's closure in 1969, the U.S. government publicly declared that it was no longer investigating UFOs. However, many researchers argue that official interest did not disappear but instead went "underground." Declassified files suggest that, while public projects ceased, classified programs continued, often under the guise of research into national security concerns. This shift in focus has led to speculation that the government may be aware of information that has not been shared with the public and might be holding evidence that would reveal more about extraterrestrial activity.

The Intelligence Community and UFOs: Documents from the CIA and NSA

While the U.S. Air Force primarily led early UFO investigations, declassified documents reveal that intelligence agencies, including the CIA and NSA, also maintained a strong interest. These agencies focused on UFO sightings within the context of Cold War security concerns, particularly when sightings coincided with sensitive military installations or aircraft testing sites. Declassified documents from these agencies provide fascinating insights into how UFO sightings were interpreted as potential espionage or military threats, prompting both detailed investigations and a cautious approach to public disclosure.

CIA Involvement and the 1953 Robertson Panel

In 1953, the CIA convened the Robertson Panel to assess the UFO phenomenon and recommend a strategy for addressing public interest. The panel concluded that UFO sightings, while not necessarily indicative of extraterrestrial activity, could threaten national security by clogging communication channels and encouraging "mass hysteria." The panel recommended a strategy of "debunking" UFO sightings to reduce public fascination, suggesting that the government actively sought to downplay UFO reports to prevent distraction from Cold War priorities.

- **Example**: The Robertson Panel recommended collaborating with prominent media outlets to ridicule and discredit UFO reports. This strategy set a precedent for how UFO information would be managed in the following decades, as declassified CIA files reveal efforts to minimize the public's focus on sightings.

NSA Monitoring of UFO Communications

The NSA's declassified UFO files reveal a different approach, focusing on monitoring communications for indications of extraterrestrial intelligence or potential threats to secure facilities. These documents show that the NSA collected reports on foreign UFO sightings and that some NSA personnel were involved in analyzing unexplained radio signals. While the majority of these signals were ultimately attributed to natural sources or human activity, the NSA's monitoring illustrates a high level of caution and curiosity within the intelligence community regarding unexplained phenomena.

- **Case Study**: One notable NSA document, titled *In Camera Affidavit*, discusses sensitive material that the NSA argued could not be released in full due to national security concerns. Although heavily redacted, this document highlights the agency's interest in maintaining confidentiality about some aspects of UFO monitoring, further fueling speculation about hidden knowledge.

Recent Developments: The Pentagon's AATIP Program and UAP Task Force

In recent years, the existence of the Advanced Aerospace Threat Identification Program (AATIP) and the establishment of the UAP (Unidentified Aerial Phenomena) Task Force have revived interest in government documentation of UFOs. AATIP, initially a secret program within the Department of Defense, focused on studying unexplained aerial phenomena to determine if they posed potential security threats. Although officially disbanded in 2012, AATIP was followed by the creation of the UAP Task Force, which continues to analyze recent encounters reported by military personnel.

AATIP's Goals and Revelations

AATIP's focus was on evaluating UAP sightings from a scientific and defense perspective. According to declassified reports and leaked documents, AATIP researchers studied radar, video, and sensor data, as well as interviews with pilots who had direct encounters with UAPs. Some cases involved objects that performed maneuvers beyond known human technology, such as rapid acceleration and directional changes without visible means of propulsion. These characteristics fueled speculation that AATIP encountered evidence of technology from an unknown source.

- **Example**: The "Tic Tac" UFO, captured on video by U.S. Navy pilots in 2004, became one of the most significant pieces of evidence examined by AATIP. The object's extraordinary speed, abrupt directional shifts, and lack of heat signature raised questions about its origin. AATIP's analysis contributed to a shift in how the government approaches UAPs, with greater openness to acknowledging the phenomenon's unknown aspects.

The Role of the UAP Task Force and Ongoing Investigations

The UAP Task Force, officially established in 2020, has continued the work of AATIP, emphasizing transparency in reporting encounters. Its formation signaled a shift in government policy, as officials released unclassified videos and provided reports to Congress. The Task Force aims to create a standardized approach for documenting and analyzing UAP encounters, particularly those involving military assets, to better understand the potential implications for national security. The increased transparency in UAP investigations has led to a broader acceptance of the phenomenon and has encouraged public interest in government disclosure.

The Challenges of Secrecy and Public Trust: What Declassified Documents Reveal

Despite the release of these documents, government transparency remains limited. Many documents are heavily redacted, leaving key details obscured. The continued withholding of information has fueled public distrust, with theories that the government possesses evidence of extraterrestrial contact or advanced technology. Examining these declassified documents, it becomes clear that official investigations have been extensive, and that the phenomenon has been taken seriously at the highest levels of government. However, the incomplete nature of these records leaves room for speculation and raises questions about the motivations for continued secrecy.

Balancing National Security and Public Knowledge

National security is often cited as the primary reason for withholding information. Some documents suggest that revealing details about UAP encounters might compromise defense technology or intelligence methods, potentially giving adversaries valuable information. This justification, while understandable, does not fully explain why so many UFO-related documents are still classified or heavily redacted, suggesting there may be additional motives for limiting public knowledge.

The Role of Public Pressure and Transparency Initiatives

In recent years, increased public demand for government transparency has led to gradual changes in disclosure policy. Organizations like The Freedom of Information Act (FOIA) and advocacy groups have pressured agencies to release previously classified UFO-related documents. The gradual release of AATIP reports, UAP Task Force findings, and older documents from the CIA and NSA reflects a cautious but growing willingness to engage with the public on this issue.

Declassified Documents and Their Role in Understanding Government Involvement

Declassified documents offer a vital yet incomplete view of governmental involvement in UFO research. Programs like Project Blue Book, the CIA's Robertson Panel, and AATIP reveal that the government has consistently maintained an interest in UFOs, though often with a focus on security and secrecy. These documents show that official concern about UFOs and UAPs extends beyond public curiosity, touching on issues of national defense, intelligence gathering, and technology monitoring.

While declassified documents provide valuable insights, the limitations of these records—often due to redactions and withheld details—mean that much remains unknown. The release of recent findings from the UAP Task Force represents a potential turning point in transparency, suggesting a shift toward greater openness. However, the history of government involvement in UFO research remains marked by a tension between secrecy and disclosure, leaving the public to question whether more knowledge is still hidden and what revelations might lie behind classified files.

Secrets in the Shadows: Exploring Theories of Government Knowledge and UFO Secrecy

For decades, the theory that governments around the world possess secret knowledge of extraterrestrial life and advanced technology has captivated the public imagination. Often labeled as conspiracy theories, these beliefs stem from a combination of government actions, such as document redactions, restricted access to information, and occasional leaks or "whistleblower" testimonies. At the heart of this topic lies a profound question: if governments indeed possess knowledge of extraterrestrial life or advanced non-human technology, why would they keep it secret? Theories about government knowledge and secrecy range from national security concerns to speculative ideas about societal impact and control, each adding layers of complexity to the study of UFO phenomena.

The Roswell Incident: The Root of Secrecy Theories

The 1947 Roswell Incident stands as a cornerstone in theories of government secrecy surrounding UFOs. After an unidentified object crashed near Roswell, New Mexico, military officials initially described it as a "flying disc." However, within days, they retracted this statement, instead labeling it a weather balloon. This abrupt shift led to widespread skepticism and laid the foundation for theories that the U.S. government was concealing the truth. Witnesses later claimed to have seen strange materials and even alien bodies at the crash site, adding further intrigue.

Roswell's Legacy and the Birth of UFO Secrecy Culture

The Roswell Incident left an indelible mark on public consciousness, creating an association between UFO sightings and government cover-ups. This event is often viewed as the starting point for the government's supposed "policy of secrecy" on extraterrestrial matters. Over the years, the Roswell story evolved into a rich tapestry of speculation, with allegations of recovered alien technology, reverse engineering, and concealed extraterrestrial bodies fueling theories about a hidden extraterrestrial presence.

- **Case Study**: In 1994, the U.S. Air Force released a report titled *The Roswell Report: Fact vs. Fiction in the New Mexico Desert*, which claimed that the object recovered in 1947 was part of Project Mogul, a classified program aimed at detecting Soviet nuclear tests. However, many UFO enthusiasts remain unconvinced, viewing the report as a continuation of government obfuscation.

National Security and Strategic Concerns: Justifying Secrecy

One of the more practical theories behind UFO secrecy is the potential threat these phenomena might pose to national security. If unidentified flying objects are indeed advanced craft capable of outperforming human technology, then their existence would raise concerns about vulnerabilities within air defense systems. Government agencies may withhold information to prevent adversaries from understanding the full extent of these security challenges or from gaining access to advanced technology through potential leaks.

The Role of Military and Intelligence Agencies

The involvement of agencies such as the CIA, NSA, and military branches in UFO investigations suggests a strategic interest that goes beyond mere curiosity. Declassified documents reveal that during the Cold War, U.S. officials feared that the Soviet Union might leverage UFO sightings to test U.S. air defenses or even disguise their own aircraft as UFOs. Given this context, secrecy becomes a rational response to potential exploitation by foreign adversaries.

- **Example**: The 1953 CIA-led Robertson Panel, convened to address public interest in UFOs, concluded that the phenomenon could be used by enemies to manipulate or overwhelm the public. As a result, the panel recommended downplaying and debunking UFO sightings as a matter of psychological security, thus initiating an official strategy of "debunking" unexplained phenomena.

Technology Protection and Reverse Engineering

Some theories propose that governments, particularly the U.S., may have recovered extraterrestrial technology through events like Roswell. By reverse-engineering these artifacts, they could potentially gain technological advantages in fields such as propulsion, stealth, and energy systems. Keeping such research classified would prevent adversaries from accessing potentially revolutionary technology. Although no concrete evidence exists to prove that extraterrestrial technology has been recovered, this theory persists due to frequent mentions of "reverse-engineering" programs in leaked documents and whistleblower testimonies.

Societal Impact: Protecting Against Mass Panic and Disruption

Another theory suggests that the government's primary motivation for UFO secrecy is to avoid societal upheaval. The notion that extraterrestrial beings have been visiting Earth raises profound questions about humanity's place in the universe, potentially disrupting religious, philosophical, and cultural worldviews. In this perspective, secrecy is seen as a means to protect societal stability by managing the disclosure process carefully, avoiding sudden shocks that could lead to widespread fear or existential crises.

Psychological Theories of Disclosure

Studies have shown that individuals respond in diverse ways to the idea of extraterrestrial life, with reactions ranging from excitement to fear and denial. Sociologists and psychologists suggest that abrupt disclosure of extraterrestrial contact could lead to unpredictable behavior, particularly among populations whose worldviews are deeply tied to anthropocentric or religious beliefs. Governments might therefore prefer a slow, controlled release of information to allow for societal adaptation.

- **Example**: In 1960, the Brookings Report, commissioned by NASA, advised that the discovery of extraterrestrial life could disrupt social systems and recommended that any potential information be carefully managed. Though not directly related to UFOs, this report influenced the mindset of many within government agencies, suggesting a precedent for secrecy to avoid social instability.

Maintaining Control and Managing Information

Another perspective posits that governments maintain UFO secrecy as a means of information control. Knowledge of extraterrestrial life would represent one of the most powerful discoveries in human history, conferring significant influence on those who possess it. From this perspective, secrecy allows governments to retain control over the narrative and manage public perception. By controlling the flow of information, officials can avoid having to answer difficult questions about human vulnerability, the limits of our scientific knowledge, and the potential limits of governmental power.

Alleged Whistleblowers and Leaked Documents: Theories from Insiders

Over the years, several individuals claiming insider knowledge have come forward with allegations that the government is concealing extraterrestrial information. Figures like Bob Lazar, who claimed to have worked on reverse-engineering extraterrestrial craft at a site near Area 51, and former intelligence officer Luis Elizondo, who led the Advanced Aerospace Threat Identification Program (AATIP), have fueled theories that knowledge of extraterrestrial encounters extends into the highest echelons of government.

Whistleblower Testimonies and Public Perception

Whistleblower accounts often align with broader theories of secrecy, suggesting that the government possesses physical evidence of extraterrestrial technology or has conducted extensive studies on alien craft. While skeptics argue that many whistleblower claims lack verifiable evidence, these stories resonate with the public, fueling the belief that certain truths remain hidden. Elizondo's involvement, particularly, lent credibility to claims that the Pentagon has taken UAPs seriously and has collected data on their behavior and potential threat.

- **Case Study**: Bob Lazar's 1989 claim of working at "S-4" near Area 51, where he allegedly encountered alien craft, created a media sensation and entrenched Area 51 as a focal point for government secrecy theories. While Lazar's background has been questioned, his story persists in UFO lore, raising questions about the government's interest in advanced, unknown technology.

Leaked Documents and Media Reports

Leaked documents, such as those released by the CIA, FBI, and other intelligence agencies, add weight to theories of secrecy by revealing the existence of classified investigations, some of which date back decades. Although often heavily redacted, these documents confirm that the government monitored UFO sightings and took reports seriously enough to document them. Media reports on programs like AATIP have further confirmed ongoing interest, even if details remain sparse.

- **Example**: In 2017, The New York Times published an article on the Pentagon's AATIP program, revealing that the Department of Defense had spent millions investigating UAPs. This report served as a partial disclosure, suggesting that interest in UFOs had continued even after the formal conclusion of Project Blue Book, fueling speculation about hidden agendas.

Extraterrestrial Diplomacy and Speculative Alliances: Outlandish but Persistent Theories

Some of the more speculative theories suggest that governments, particularly superpowers, might be in direct communication with extraterrestrial civilizations. In these theories, secrecy is maintained as part of an interstellar diplomatic agreement, with extraterrestrial beings either demanding secrecy or cooperating with governments in exchange for technology sharing or other resources. While lacking credible evidence, this idea remains popular within certain circles of UFO researchers and adds an element of intrigue to the broader theories about secrecy.

The Concept of "Galactic Agreements"

The notion of a secret agreement between governments and extraterrestrial beings is often referred to as a "Galactic Agreement" or "treaty." Proponents argue that extraterrestrial civilizations, recognizing the potential disruption their existence could cause, might impose conditions on contact with humanity, including secrecy and controlled disclosures. Though this theory lacks verifiable support, it serves as a compelling narrative that fits within larger theories about government cover-ups and hidden extraterrestrial alliances.

Deciphering the Motivations Behind Government Secrecy

Theories of government knowledge and secrecy regarding UFOs and extraterrestrial encounters reflect a range of plausible and speculative motivations. From national security concerns to societal impact, technological protection, and information control, these theories underscore the complex factors that may contribute to official silence on unexplained phenomena. Incidents like the Roswell crash, historical reports from the CIA and military branches, and recent developments through AATIP illustrate an undeniable pattern of governmental interest, though the depth and nature of that interest remain largely unknown.

While speculative theories such as "Galactic Agreements" add an element of fascination, practical concerns like protecting national security, avoiding public panic, and preserving technological superiority provide more grounded explanations for secrecy. Whistleblower testimonies and media disclosures add weight to the argument that knowledge of UFOs extends beyond public purview, but the lack of comprehensive evidence leaves many questions unanswered.

In the end, theories about government secrecy on UFOs reveal more about human curiosity, cultural anxieties, and the desire to understand our place in the universe. Whether driven by necessity or by design, government silence on UFOs has created a fertile ground for theories, each contributing to a complex and layered understanding of how humans interpret the unknown.

Summary of Evaluating the Credibility and Authenticity of Messages

In examining the credibility and authenticity of extraterrestrial messages, we've traversed multiple dimensions that influence how society perceives these phenomena. We began by assessing witness credibility, exploring both the backgrounds of individuals who report encounters and the importance of corroborating their stories with objective observations. Psychological and social factors also shape credibility, as personal beliefs, cultural influences, and even psychological conditions can alter perception, memory, and interpretation. These insights emphasize the need for a nuanced approach to witness testimony, as credibility can vary widely based on individual and situational factors.

Our exploration of physical evidence and documentation revealed both the value and limitations of material proof in extraterrestrial encounters. Physical evidence, such as soil samples and radar data, can lend scientific weight to reports, yet challenges persist in verifying and preserving such evidence. Similarly, while photographic and video evidence offers visual documentation, advancements in technology and the potential for manipulation demand a cautious approach to analysis. Here, understanding the limitations and vulnerabilities of visual evidence is as crucial as the evidence itself.

Finally, we examined government involvement and cover-ups, a topic that sits at the intersection of secrecy, public speculation, and the broader quest for knowledge. Analysis of declassified documents provided insight into historical and modern government responses to UFO phenomena, illustrating how defense agencies have often treated these encounters as potential national security issues. Furthermore, theories of government knowledge and secrecy, from national security concerns to speculative ideas about extraterrestrial alliances, offer compelling narratives that continue to intrigue the public. These theories underscore the delicate balance between transparency, control, and the potential impact of such knowledge on society.

Together, these topics underscore the complexity of assessing the credibility and authenticity of messages related to extraterrestrial encounters. Each layer—from witness testimony and physical evidence to government secrecy—adds depth to our understanding, highlighting that while mystery surrounds the phenomena, a rigorous approach helps distinguish credible elements from speculation. This evaluation of credibility and authenticity provides a critical foundation for moving forward with informed, open-minded analysis as we continue exploring the mysteries that may define humanity's place in the cosmos.

Having carefully evaluated the credibility and authenticity of extraterrestrial messages, we now turn to a question that strikes at the heart of these encounters: what relevance might these messages hold for humanity and our future? If genuine, these communications offer profound insights, addressing urgent global concerns like environmental degradation, nuclear proliferation, and the evolution of human consciousness. By examining recurring themes—such as environmental stewardship, nuclear disarmament, and spiritual growth—we gain a better understanding of the potential guidance embedded within these messages. This exploration challenges us not only to consider the origins of these warnings but also to contemplate how humanity might act in response. As we delve into these messages, we will investigate both the practical actions and philosophical shifts they may inspire, ultimately asking ourselves whether we are prepared to heed the call for a more harmonious and conscientious future.

Chapter 9: Messages for a Sustainable and Peaceful Future: Humanity's Responsibility in Extraterrestrial Warnings

Throughout the study of extraterrestrial messages, certain themes emerge that resonate deeply with global challenges humanity faces today. These messages often touch on urgent topics like environmental sustainability, the risks of nuclear weapons, and the potential for spiritual evolution. If these communications are indeed attempts to guide humanity, they may hold crucial insights on how we might address issues that threaten our collective future. From stark warnings about environmental destruction and the catastrophic consequences of nuclear warfare to encouragements for spiritual growth and inner peace, these messages invite us to reconsider our relationship with the planet, each other, and ourselves. In this section, we will explore the content and implications of these messages, looking at both the practical steps they may suggest and the deeper transformations they may inspire. Are these communications urging us to adopt a new level of stewardship and responsibility? Through careful analysis, we can uncover the wisdom these messages may offer for building a sustainable, peaceful, and spiritually aware future.

Guardians of the Earth: Examining Environmental Warnings in Extraterrestrial Messages

Over the decades, numerous witnesses of extraterrestrial encounters have reported receiving messages warning of environmental dangers. The consistency of these environmental themes across encounters raises intriguing questions: if extraterrestrial beings are truly trying to communicate with us, why would they express such urgent concern for Earth's ecological wellbeing? Many messages seem to point to pollution, climate change, deforestation, and species extinction as critical issues threatening not only humanity but also the planet as a whole. These messages often paint a vivid picture of a future in which humanity's negligence has led to a deteriorating environment, encouraging us to take responsibility as stewards of the Earth. This exploration seeks to dive deep into these recurring environmental warnings, revealing the themes, motivations, and practical guidance embedded within these messages and challenging us to reflect on our role in shaping Earth's future.

The Planet in Peril: A Call to Action from Extraterrestrial Sources

Reports of extraterrestrial messages addressing environmental issues date back as early as the 1950s, particularly within the Contactee Movement. Many contactees, such as George Adamski and Howard Menger, claimed that extraterrestrial beings voiced concerns over the deteriorating state of Earth's environment. These messages would often come in a tone of caution or gentle guidance, warning humanity about its trajectory of industrialization and environmental degradation. Over the decades, these themes became even more urgent, with more recent accounts emphasizing the immediate consequences of pollution, deforestation, and climate change.

Some witnesses describe extraterrestrial beings showing them distressing images of Earth's potential future: barren landscapes, polluted oceans, and skies filled with toxins. These visual depictions appear to serve as cautionary tales, intended to awaken humanity to the urgent need for environmental stewardship. Often, witnesses report feeling a profound sense of sorrow or responsibility upon receiving these images, suggesting that extraterrestrial beings may be appealing not just to our intellect but to our emotions and moral compass.

The Message Behind the Message: Understanding the Ecological Themes

The recurring themes in extraterrestrial messages about the environment can be broken down into a few central concerns:

1. **Pollution and Toxicity**: Many messages emphasize the damage being done to Earth's air, water, and soil due to industrial pollution. Witnesses report being warned about the build-up of toxins from human activities—waste from factories, plastic pollution, and heavy metals contaminating natural systems. These messages often highlight the direct connection between human health and environmental health, urging us to recognize the dangers of unchecked pollution.
2. **Climate Change**: As the effects of climate change became increasingly evident in the 1980s and beyond, reports of extraterrestrial messages echoing these warnings grew more frequent. Descriptions of rising global temperatures, melting ice caps, and extreme weather events appeared in multiple encounters. The beings delivering these messages often stress that humanity must act quickly to reduce greenhouse gas emissions, pointing out that continued inaction will lead to irreversible damage.
3. **Biodiversity and Habitat Loss**: Some encounters include messages that underscore the loss of biodiversity and the destruction of natural habitats. Witnesses describe being told of the interconnectedness of all species and the delicate balance that sustains ecosystems. By disrupting this balance, humans risk ecological collapse—a concept that resonates with modern environmental science,

which highlights the vital role biodiversity plays in sustaining life.
4. **Sustainable Practices**: In addition to warnings, many of these messages propose alternative ways of living that could help humanity mitigate or reverse environmental damage. Suggestions include moving away from fossil fuels, adopting sustainable agricultural practices, and embracing renewable energy sources. Some witnesses describe being shown advanced technologies or harmonious systems that extraterrestrial civilizations allegedly use to coexist with their own planetary environments.

Humanity's Responsibility: Are We the "Caretakers" of Earth?

A powerful motif in many of these messages is the idea that humanity has a unique role as the caretaker or steward of Earth. This idea suggests that, unlike extraterrestrial civilizations that may have learned to coexist sustainably with their planets, humans are still grappling with the responsibility of maintaining ecological balance. The warnings seem to imply that Earth is a rare and precious planet, with unique biodiversity and resources that require active protection. This perspective not only emphasizes the gravity of our environmental impact but also serves as a reminder of the moral and ethical responsibilities we hold toward future generations.

In several accounts, witnesses recount feeling a sense of accountability upon receiving these messages, as if extraterrestrial beings were urging them to lead by example or advocate for environmental preservation. The notion that humans are expected to evolve toward a more sustainable existence suggests that extraterrestrial beings might view environmental stewardship as an essential criterion for humanity's advancement—perhaps even a prerequisite for any deeper engagement with advanced civilizations.

A Broader Perspective on Environmental Crisis: A Shared Responsibility?

One of the more thought-provoking aspects of these messages is the implication that Earth's environmental crisis extends beyond humanity, potentially affecting other planetary systems or universal energy balances. This perspective aligns with certain Indigenous beliefs and spiritual philosophies that consider Earth a part of a larger cosmic network. If extraterrestrial beings truly view Earth's wellbeing as a matter of collective concern, then humanity's environmental choices might resonate far beyond our planet.

This concept also hints at the possibility of a shared responsibility. Some messages suggest that, just as extraterrestrial beings may have faced similar ecological challenges in their histories, humanity must now undergo its own process of environmental reckoning and realignment. In this way, the environmental themes present in extraterrestrial messages invite us to think not only about personal and collective responsibility on a planetary scale but also about our interconnectedness with other potential civilizations.

Turning Warnings into Action: Practical Guidance from the Messages

Beyond serving as warnings, many of these messages include practical suggestions that align closely with the goals of modern environmental movements. Witnesses often describe being encouraged to:

- **Reduce Reliance on Fossil Fuels**: Extraterrestrial messages frequently stress the dangers of continuing to rely on fossil fuels, citing pollution and climate change as primary concerns. They often advocate for the adoption of renewable energy sources such as solar, wind, and potentially even more advanced forms of energy that are as yet unknown to humanity.
- **Preserve Natural Ecosystems**: Witnesses report that extraterrestrial beings emphasize the importance of preserving forests, oceans, and other natural habitats. In doing so, humanity can protect biodiversity and maintain ecological balance, preventing the collapse of vital systems that support life on Earth.
- **Promote Environmental Awareness**: Many messages contain an underlying theme of education, urging humans to raise awareness of environmental issues among others. By understanding the impacts of their actions and choices, people can make informed decisions that contribute to sustainability.
- **Engage in Spiritual Practices of Connection**: Some messages go further, suggesting that humans reconnect with nature through practices such as meditation, mindfulness, and traditional forms of spirituality. This approach echoes the belief that environmental health is closely tied to human consciousness and that a shift in mindset can foster a more harmonious relationship with the planet.

The Urgent Relevance of Environmental Messages

The environmental warnings conveyed in these extraterrestrial messages offer a profound call to action. They speak not only to our physical survival but to the ethical responsibilities we hold as stewards of Earth. From pollution and climate change to biodiversity loss and the need for sustainable practices, these messages encapsulate the major environmental crises of our time, challenging us to confront the consequences of our actions and adopt a more sustainable, conscientious approach to life on this planet.

Through recurring themes and practical guidance, these messages underscore the importance of proactive environmental care, suggesting that extraterrestrial beings may view such stewardship as a fundamental quality of advanced civilizations. The sense of urgency conveyed in these messages serves as both a warning and a potential roadmap, inviting us to embrace a sustainable future. Whether extraterrestrial in origin or not, these messages offer valuable insights, reminding us of the power we hold to shape the destiny of our planet and calling us to rise to the challenge of protecting the Earth for future generations.

Taking Responsibility: Humanity's Path Forward in Response to Environmental Warnings

Extraterrestrial messages reported by individuals across decades often emphasize that humanity stands at a critical crossroads. These warnings about environmental destruction are not mere admonishments; they are calls to action, urging us to adopt sustainable practices that safeguard the Earth for future generations. The proposed actions are not only about reducing harm but also about evolving toward a harmonious relationship with the environment, guided by principles of respect, conservation, and balance. By examining these actions, we can gain insight into the transformative steps humanity could take to meet the environmental challenges these messages highlight.

Transitioning to Renewable Energy: Moving Beyond Fossil Fuels

One of the central themes of extraterrestrial messages is the urgent need to transition away from fossil fuels, which are often cited as a primary cause of pollution, climate change, and ecological disruption. Extraterrestrial warnings echo many of the same concerns that scientists and environmental advocates have raised about the unsustainable nature of fossil fuels, which not only harm the planet but are finite resources.

- **Developing Renewable Energy Technologies**: Solar, wind, and hydroelectric power offer sustainable alternatives that could reduce our reliance on fossil fuels. By investing in these technologies, governments and industries can create infrastructures that harness clean energy on a large scale. Extraterrestrial messages often allude to advanced civilizations having achieved mastery over clean energy sources, suggesting that our future lies in renewable energy.
- **Localized and Decentralized Energy**: Many messages suggest that humanity must rethink centralized energy grids, proposing localized systems that allow communities to be self-sufficient. Decentralized energy sources, such as solar panels on homes or micro-wind turbines, reduce the strain on large power plants and promote resilience.

Protecting Natural Ecosystems: Preserving the Web of Life

Extraterrestrial messages frequently highlight the interconnectedness of all life on Earth and stress the importance of protecting ecosystems. This emphasis aligns with the ecological concept that each species plays a role in maintaining the balance of natural systems. Destruction of habitats and biodiversity not only threatens individual species but also destabilizes the broader ecosystem.

- **Reforestation and Conservation Efforts**: Deforestation is a critical concern mentioned in these warnings. Restoring degraded landscapes and protecting existing forests are vital for maintaining biodiversity, preserving water cycles, and combating climate change. Reforestation initiatives, particularly in tropical regions, are crucial for carbon sequestration and biodiversity conservation.
- **Protecting Marine Ecosystems**: Oceans are a major focus in extraterrestrial messages, as they play a central role in regulating the Earth's climate and supporting life. Efforts to reduce plastic pollution, protect coral reefs, and combat overfishing are essential for preserving the health of marine ecosystems. Initiatives such as marine protected areas (MPAs) provide sanctuaries where marine life can thrive without human interference.

Reducing Pollution: Addressing the Threat of Toxins

Pollution is a recurring theme in extraterrestrial messages, particularly warnings about the impact of industrial waste, chemicals, and plastic pollution. These messages often focus on the importance of reducing our environmental footprint and preventing toxins from entering natural ecosystems.

- **Minimizing Plastic Use and Waste**: Reducing plastic production and waste is essential for reducing pollution. Policies that encourage the use of biodegradable alternatives and promote recycling can help address the significant environmental impact of plastic.
- **Reducing Chemical Runoff**: Extraterrestrial messages frequently emphasize the damage done to water sources by agricultural and industrial runoff.

Implementing sustainable agricultural practices, such as reducing pesticide use and adopting organic farming, can prevent harmful chemicals from contaminating water bodies.

Educating for Environmental Awareness: Fostering a Culture of Responsibility

A critical aspect of responding to environmental warnings is educating the public about the impact of human actions on the planet. Many extraterrestrial messages highlight the role of awareness in shaping a sustainable future, suggesting that humanity must cultivate an understanding of its interconnectedness with the environment.

- **Integrating Environmental Education**: Including environmental studies in educational curricula can create a foundation of knowledge that fosters responsible behavior. Understanding the ecological impact of human actions encourages individuals to make informed choices and advocate for sustainability.
- **Public Awareness Campaigns**: Raising awareness through campaigns, documentaries, and social media can reach a broader audience. By using storytelling techniques to convey the gravity of environmental issues, these campaigns can inspire people to take meaningful action in their daily lives.

Embracing Sustainable Practices in Daily Life

In addition to large-scale policy changes and technological advancements, extraterrestrial messages suggest that individual actions are vital in creating a sustainable future. These messages often emphasize that each person has a role in the collective effort to protect the planet.

- **Reducing Carbon Footprints**: Simple actions such as conserving energy, reducing car usage, and supporting

eco-friendly products can contribute to reducing carbon emissions. By making conscious choices, individuals can play a part in mitigating climate change.
- **Promoting Sustainable Consumption**: Extraterrestrial warnings frequently allude to the concept of sustainable consumption, encouraging humans to limit waste and adopt minimalist lifestyles. By choosing products with lower environmental impacts and supporting ethical companies, individuals can help create a market that prioritizes sustainability.

Collective Action and Global Collaboration: A Unified Response

A recurring message in extraterrestrial communications is the need for global cooperation to address environmental challenges. Environmental issues transcend national borders, requiring collaborative solutions that reflect a commitment to the planet rather than individual nations' interests.

- **International Environmental Treaties**: Global treaties, such as the Paris Agreement on climate change, represent an essential step toward coordinated environmental action. Extraterrestrial messages often emphasize that humanity must unite to protect Earth, transcending geopolitical divisions.
- **Building Resilient Communities**: Collective action also involves empowering local communities to respond to environmental issues. By fostering resilience and supporting sustainable development, societies can reduce vulnerability to environmental crises.

Translating Warnings into Action

In response to environmental warnings from extraterrestrial messages, humanity is presented with a profound opportunity to reshape its relationship with the planet. The proposed actions align with the goals of modern environmentalism, urging a transition to renewable energy, the preservation of ecosystems, pollution reduction, and public education on sustainability. These messages suggest that true progress requires both individual and collective responsibility, as well as a shift in consciousness that places environmental stewardship at the core of human values.

By heeding these warnings and embracing sustainable practices, humanity can not only avert environmental catastrophe but also evolve toward a harmonious, resilient future. This path represents more than just environmental reform; it symbolizes a transformation in how we view our place within Earth's ecosystems, challenging us to act as guardians rather than mere inhabitants of the planet. Through meaningful action, humanity can demonstrate its readiness to assume the role of responsible custodians, ensuring that Earth remains a vibrant home for generations to come.

Turning the Tide: The Call for Nuclear Disarmament and Global Peace Initiatives

Nuclear disarmament stands as one of the most pressing themes in reported extraterrestrial messages, reflecting an urgent call for humanity to abandon weapons capable of catastrophic destruction. Many encounters with purported extraterrestrial beings reveal concerns about humanity's reliance on nuclear weapons, warning of their devastating potential and urging a path toward disarmament. The emphasis on nuclear disarmament is not only a call to avoid physical destruction but also a message advocating for a shift in global consciousness—from one rooted in power and conflict to one that embraces peace and cooperation. Understanding the relevance of these warnings to global disarmament efforts provides a lens through which we can examine the broader implications of extraterrestrial messages, delving into how humanity might rise to the challenge of forging a world free from the threat of nuclear war.

The Origin of Nuclear Warnings in Extraterrestrial Messages

The appearance of nuclear-related themes in reported extraterrestrial messages began shortly after the first use of nuclear weapons during World War II. As humanity entered the Atomic Age with the bombings of Hiroshima and Nagasaki, reports of UFO sightings and alleged extraterrestrial encounters increased. Many witnesses have described receiving warnings about the dangers of nuclear technology, with extraterrestrial beings reportedly expressing concern over the destructive potential of these weapons. These messages often emphasize the far-reaching consequences of nuclear warfare, not only for human civilizations but also for the broader balance of life on Earth and beyond.

Extraterrestrial warnings about nuclear weapons tend to include references to both their immediate destructive capacity and their long-term ecological impact. Witnesses have recounted that extraterrestrial beings caution humanity about the environmental damage and radiation fallout that would accompany nuclear conflict, with the potential to render entire regions uninhabitable. These warnings suggest that extraterrestrial beings view the use of nuclear technology as incompatible with humanity's potential for growth and evolution, encouraging a shift away from militaristic pursuits.

The Global Disarmament Imperative: Moving Beyond Nuclear Conflict

One of the core messages in these nuclear warnings is a call for disarmament—a concept that aligns with global disarmament efforts aimed at reducing the nuclear threat. International treaties, like the Treaty on the Non-Proliferation of Nuclear Weapons (NPT) and the recent Treaty on the Prohibition of Nuclear Weapons, reflect humanity's understanding of the existential risk posed by nuclear arms. Extraterrestrial messages, however, often push the concept of disarmament further, urging complete abolition of nuclear weapons and an end to their development altogether.

The Ethical and Moral Dimensions of Disarmament

Many extraterrestrial messages suggest that nuclear disarmament is not only a practical necessity but a moral imperative. These warnings portray nuclear weapons as fundamentally incompatible with peace, calling for a reassessment of humanity's ethical stance on weapons of mass destruction. By abandoning nuclear arms, humanity would take a step toward embodying values such as compassion, respect for life, and cooperation—qualities that many extraterrestrial messages imply are necessary for a harmonious existence.

This ethical dimension reflects the idea that nuclear disarmament is more than a political act; it is a transformational shift in humanity's collective mindset. Extraterrestrial beings are often described as presenting disarmament as a necessary step in humanity's journey toward a more enlightened, peaceful state, one in which conflicts are resolved through diplomacy rather than through threats of annihilation.

The Potential Consequences of Ignoring Nuclear Warnings

Reports of extraterrestrial messages often include dire predictions about the potential consequences of nuclear conflict. Witnesses describe vivid visions or warnings that illustrate the devastation nuclear warfare could unleash, not only on human populations but on the environment and future generations. These warnings frequently reference the lasting impact of radiation on ecosystems, illustrating the severe, long-term consequences of nuclear explosions.

- **Human and Ecological Impact**: Messages typically emphasize the wide-reaching effects of nuclear explosions, including fallout, loss of life, genetic mutations, and contamination of water and soil. Witnesses often describe a sense of horror and responsibility upon seeing these visions, reflecting the profound consequences of nuclear actions.
- **Interference with Planetary Balance**: Some extraterrestrial messages imply that the use of nuclear weapons could disrupt the natural energy balances of the planet, potentially causing repercussions that extend beyond Earth. This concept suggests that extraterrestrial civilizations view nuclear weapons as a threat not only to humanity but also to other planetary systems or interstellar networks, adding a cosmic dimension to the urgency of disarmament.

Encouraging a New Mindset: Fostering Peaceful Relations

In addition to nuclear disarmament, extraterrestrial messages often advocate for a broader commitment to peace and cooperation. These messages encourage humanity to reject violence as a means of resolving conflicts, instead embracing diplomacy and mutual understanding. This shift aligns with the principles promoted by international peace organizations and leaders, who argue that genuine security can only be achieved through cooperative, rather than confrontational, relationships.

Reimagining Security Through Peace

Many of these messages propose a redefinition of what it means to be "secure." Instead of relying on weapons of mass destruction, security could be based on the strength of alliances, open communication, and mutual respect. This view aligns with the goals of peace studies and conflict resolution fields, which emphasize the importance of dialogue and compromise over coercive tactics. By redefining security in this way, humanity would not only protect itself from the dangers of nuclear warfare but also foster a global environment that supports peace.

- **Promoting International Cooperation**: Extraterrestrial messages often underscore the need for humanity to come together as a unified species, transcending national and cultural boundaries. In doing so, humans would be able to prioritize collective survival and well-being over individual national interests, aligning with principles of internationalism and shared responsibility.
- **Embracing Diplomatic Solutions**: These messages also highlight the need to pursue diplomatic avenues in resolving conflicts. By cultivating skills such as negotiation, active listening, and empathy, humanity can shift away from aggressive tactics and towards constructive dialogue. This approach not only reduces the risk of nuclear confrontation but also fosters a more inclusive, peaceful world order.

Lessons from Extraterrestrial Civilizations: A Model for Peace

Some reported messages suggest that extraterrestrial civilizations themselves have undergone similar processes of disarmament and pacification. Witnesses describe extraterrestrial beings who warn humanity of the paths they themselves once took, filled with destructive wars before eventually embracing peaceful coexistence. This perspective implies that more advanced civilizations have learned the hard lessons of nuclear or similar warfare and now seek to guide humanity away from the mistakes of the past.

This potential guidance suggests that humanity, too, could evolve past the need for nuclear weapons and adopt more enlightened forms of conflict resolution. Extraterrestrial messages often imply that peaceful coexistence is a mark of advanced civilization, suggesting that, for humanity to join a "cosmic community," it must first embrace peace and reject violence.

The Path Forward: Steps Toward Global Disarmament

In line with these warnings, there are several practical steps humanity could take toward nuclear disarmament and global peace initiatives:

- **Supporting and Expanding Disarmament Treaties**: International treaties that limit or eliminate nuclear weapons are a key pathway to global disarmament. By supporting treaties like the NPT and advocating for the ratification of the Treaty on the Prohibition of Nuclear Weapons, nations can move toward a shared commitment to nonproliferation.
- **Fostering a Global Culture of Peace**: Education plays a critical role in shifting mindsets away from militarism. Implementing peace studies in schools, promoting awareness of nuclear dangers, and encouraging nonviolent communication can foster a culture that values diplomatic solutions over aggression.
- **Investing in Peacebuilding and Conflict Resolution**: Governments and organizations can prioritize funding

for peacebuilding initiatives, such as mediation and international cooperation programs. By empowering communities to resolve conflicts without violence, humanity can lay the groundwork for a future free from nuclear threats.
- **Encouraging Citizen Advocacy**: Individuals have the power to influence policy by advocating for disarmament and supporting leaders who prioritize peace. Grassroots movements, petitions, and peaceful protests can amplify public opposition to nuclear weapons, applying pressure on governments to pursue disarmament.

Summary: Heeding the Call for a Peaceful Future

The extraterrestrial messages related to nuclear disarmament resonate deeply with humanity's own growing awareness of the dangers posed by nuclear technology. They call for a transformative shift in how we approach conflict and security, urging humanity to abandon weapons of mass destruction and embrace a future grounded in peace and cooperation. These messages suggest that disarmament is not just about preventing destruction but also about realigning our collective values toward a more enlightened and ethical existence.

By taking meaningful steps toward nuclear disarmament, humanity would not only reduce the immediate threat of nuclear war but also demonstrate its readiness to evolve as a civilization. These messages challenge us to envision a world in which security is rooted in mutual respect rather than in fear, and they remind us that our choices today shape the legacy we leave for future generations. Whether seen as warnings or as guides to a better future, these messages encourage humanity to rise above division and conflict, united in the pursuit of a peaceful world.

Navigating the Political Implications of Extraterrestrial Nuclear Warnings: A Call for Global Peace and Responsibility

When extraterrestrial messages emphasize the dangers of nuclear weapons, the political implications reach far beyond mere caution—they resonate with the complex realities of international relations, national sovereignty, and global security. These warnings challenge existing power structures, prompt questions about humanity's readiness for peace, and highlight the necessity for new frameworks of cooperation. The reported warnings have a unique influence on political thought, as they suggest that the global nuclear issue extends beyond human borders, impacting a broader, possibly cosmic, community. Analyzing these political implications offers a perspective on how extraterrestrial concerns could inspire governments to rethink policies on nuclear arms and prioritize international peace.

Extraterrestrial Messages as Political Disruptors: A New Paradigm of Accountability

The political landscape surrounding nuclear weapons is deeply rooted in national interests, strategic alliances, and deterrence theories. Yet, the extraterrestrial warnings introduce a disruptive element, one that reframes nuclear disarmament as a moral imperative transcending Earthly politics. Messages from purported extraterrestrial sources often imply that humanity's handling of nuclear technology affects not only Earth but the broader universe, challenging the legitimacy of any nation's sole authority over nuclear arms.

- **Redefining National Sovereignty**: Extraterrestrial warnings question whether any one country or group of countries has the right to possess or deploy nuclear

weapons, given the potential for global or even universal impact. This perspective urges leaders to consider nuclear technology as a shared responsibility and suggests that traditional notions of sovereignty are inadequate in the face of such existential threats.
- **A Call for Global Leadership in Disarmament**: These messages seem to suggest that true leadership requires a commitment to disarmament. By heeding extraterrestrial advice, nations could assume a global moral leadership role, setting an example in abandoning the pursuit of nuclear dominance. Countries that take steps toward disarmament could earn respect not only domestically but also on the international stage, reinforcing the political and ethical authority of those advocating for peace.

The Challenge to Nuclear Deterrence: A Shift in Security Doctrine

One of the most profound implications of extraterrestrial nuclear warnings is the challenge they present to the doctrine of nuclear deterrence. For decades, nations have justified nuclear arsenals as necessary deterrents against aggression, maintaining that "mutually assured destruction" (MAD) prevents large-scale wars. However, extraterrestrial messages suggest that reliance on such a doctrine is short-sighted, warning that the very existence of these weapons endangers humanity's future.

- **Questioning the Logic of Deterrence**: Extraterrestrial warnings imply that a security doctrine based on fear and annihilation is unsustainable. They encourage a shift away from deterrence and toward a peace-based doctrine that prioritizes diplomacy over the threat of force. Politically, this would require governments to deconstruct longstanding military strategies and develop new frameworks for achieving security through cooperation rather than competition.

- **Encouraging Demilitarization and Peacebuilding**: By promoting the abandonment of deterrence as a guiding principle, these messages also inspire demilitarization efforts. Extraterrestrial sources often describe advanced civilizations as ones that have transcended the need for weapons, suggesting that true advancement is marked by a commitment to peacebuilding and nonviolent conflict resolution. This shift could impact defense budgets, political alliances, and the global arms industry, redirecting resources toward peaceful initiatives.

Diplomatic Implications: Redefining Alliances and Global Governance

Extraterrestrial nuclear warnings invite nations to reconsider their political alliances and encourage collective responsibility for disarmament. If these warnings are to be taken seriously, they demand international cooperation that transcends existing rivalries, challenging countries to prioritize global stability over competitive advantage.

- **Strengthening Global Institutions**: To respond to extraterrestrial warnings effectively, global institutions like the United Nations would need to be strengthened and empowered to enforce nuclear disarmament measures. This approach implies a shift from nationalistic policies toward a form of global governance where international bodies can mediate nuclear policies and ensure compliance. It also raises questions about the role and influence of existing power structures and how they might adapt to a collective security model.
- **Redefining International Alliances**: Extraterrestrial messages could prompt nations to reconsider alliances based on military strength and instead prioritize partnerships grounded in shared commitments to peace. Traditional alliances, often centered on defense, could evolve into cooperative frameworks focused on

disarmament and mutual aid. Such alliances would promote peace initiatives, humanitarian goals, and environmental responsibility, creating a new global network dedicated to the shared objective of planetary survival.

Moral and Ethical Leadership: Inspiring New Political Standards

At the heart of the extraterrestrial nuclear warnings lies an ethical appeal, suggesting that humanity must rise to a new standard of moral responsibility. Politicians and leaders are called to set aside short-term interests and focus on the long-term wellbeing of humanity. These warnings challenge leaders to adopt a more ethical approach to decision-making, one that considers the consequences of nuclear policies for future generations.

- **Elevating Peace as a Political Priority**: The emphasis on disarmament in extraterrestrial messages implies that peace should be a central priority for policymakers. This shift calls for a transformation in political rhetoric and action, where promoting peace is as critical as economic development or national security. Leaders who prioritize peace could shape a new political paradigm, one that appeals to an increasingly conscientious global citizenry.
- **Influencing Public Opinion and Cultural Norms**: The political implications of these messages extend to how societies view nuclear weapons and military power. Extraterrestrial warnings encourage public engagement in disarmament efforts, fostering a cultural shift where citizens hold governments accountable for their nuclear policies. If citizens are motivated to demand disarmament, political leaders may be compelled to respond, gradually shifting the political climate toward one that values peace and opposes the use of nuclear arms.

The Strategic Shift: Aligning Policy with a Global Responsibility

In response to extraterrestrial nuclear warnings, some governments could strategically position themselves as champions of disarmament, thereby gaining moral influence on the world stage. This approach aligns with the idea that disarmament is not merely a humanitarian goal but a pragmatic strategy for future cooperation with potential extraterrestrial civilizations.

- **Advocating for a Peace-Based International Order**: Countries that align their policies with the peace-oriented messages from extraterrestrial sources could lead by example, promoting policies that reflect a commitment to nonviolence. This approach would require active participation in disarmament treaties, a reduction in military expenditures, and an emphasis on conflict prevention through diplomacy.
- **Promoting Ethical Influence Over Military Power**: The warnings also suggest that nations might gain more influence through ethical leadership rather than military dominance. By positioning themselves as advocates for peace and disarmament, countries could build alliances based on shared values rather than power. This influence might also prepare humanity for possible future interactions with extraterrestrial civilizations, demonstrating that Earth's societies are capable of prioritizing ethical considerations over militaristic ones.

Extraterrestrial Warnings as a Catalyst for a Peaceful Political Revolution

The overarching political implication of these extraterrestrial nuclear warnings is a potential paradigm shift in global governance—a movement toward unity, collaboration, and peace as central tenets of international relations. By embracing these messages, governments could transform political priorities, redirecting resources from military power to peacebuilding and environmental protection. Such a shift could redefine the political landscape, creating a world where cooperation and mutual responsibility replace competition and conflict.

Extraterrestrial messages emphasize that the fate of humanity lies in its ability to unite across borders, placing the wellbeing of Earth above the interests of individual nations. This perspective calls for unprecedented political courage and visionary leadership, encouraging politicians and citizens alike to imagine a future where nuclear weapons are obsolete, and peace is the foundational principle of governance.

The Political Implications of Extraterrestrial Nuclear Warnings

The political implications of extraterrestrial nuclear warnings are profound, challenging humanity to rethink the fundamental structures of international relations and security. These warnings prompt us to consider nuclear disarmament not just as a strategic choice but as a moral responsibility, one that demands global cooperation and a reimagining of sovereignty. They call into question the traditional reliance on nuclear deterrence, urging leaders to adopt a peace-based security doctrine that prioritizes diplomacy over fear.

In response to these warnings, humanity is invited to build a world where security is achieved through alliances based on shared values rather than military might, and where ethical considerations shape political decisions. These messages highlight the potential for political transformation, inspiring leaders to champion disarmament and embody a new era of peace-centered governance. Through the lens of extraterrestrial warnings, humanity's political journey becomes not only a quest for survival but also a pathway to a more unified and enlightened global society.

Awakening the Human Spirit: Extraterrestrial Messages as a Call for Spiritual and Consciousness Evolution

Many reported encounters with extraterrestrial beings contain messages that emphasize the importance of spiritual growth and consciousness evolution. Unlike the warnings about nuclear or environmental threats, which are often grounded in tangible actions and immediate consequences, these messages speak to a more abstract yet profoundly impactful dimension of human existence. Through urging humanity to reach higher levels of consciousness and spiritual awareness, these messages challenge us to go beyond our materialistic pursuits and cultivate qualities of compassion, unity, and wisdom. Understanding these messages requires exploring how extraterrestrial beings might view spiritual growth, why they see it as essential for humanity's evolution, and how individuals and societies can respond to this call for awakening.

The Spiritual Dimension in Extraterrestrial Messages: A New Pathway to Human Evolution

Reports of extraterrestrial encounters often describe beings that embody qualities associated with advanced spiritual awareness—radiating calm, wisdom, or an aura of peace. Many witnesses describe feeling an overwhelming sense of love, interconnectedness, or enlightenment during these encounters, suggesting that extraterrestrials are not only technologically advanced but also spiritually evolved. These experiences hint that spiritual evolution is not a secondary concern but rather an integral part of advancing as a civilization.

- **A Model of Advanced Spirituality**: In these messages, extraterrestrial beings frequently model behaviors that reflect high levels of empathy, respect, and unity. Their approach suggests that genuine progress as a civilization includes nurturing spiritual qualities, indicating that they consider these traits essential for peaceful coexistence. The beings often emphasize values like nonviolence, mutual respect, and interdependence—qualities that many humans might associate with enlightenment or spiritual maturity.
- **Spirituality as the Foundation of Advancement**: Extraterrestrial messages suggest that humanity's technological advancements are only part of the equation for true progress. According to these accounts, spiritual evolution is as crucial as scientific and intellectual achievements, creating a balanced foundation for a thriving society. By adopting this perspective, humanity is invited to shift from a predominantly technological focus toward one that honors spiritual development as the guiding force in its evolution.

Encouraging Personal Transformation: The Inner Path to Awakening

At the heart of many extraterrestrial messages is an appeal for individuals to embark on personal spiritual journeys. These messages often emphasize the importance of inner transformation, suggesting that humanity's collective growth begins with the self. By fostering qualities such as compassion, mindfulness, and self-awareness, individuals can contribute to the evolution of human consciousness on a global scale.

- **Practicing Compassion and Empathy**: Extraterrestrial messages often stress compassion as a core tenet of spiritual growth. Individuals are encouraged to cultivate empathy for others, fostering a sense of interconnectedness that transcends cultural, national, and species boundaries. By recognizing the inherent worth of all life, people can begin to embody the unity that extraterrestrial messages suggest is vital for humanity's evolution.

- **Developing Mindfulness and Presence**: Many accounts describe extraterrestrial beings as deeply mindful, fully present in the moment, and connected to their surroundings. Emulating these qualities can lead to greater self-awareness and clarity, allowing individuals to recognize and overcome personal limitations or biases. Mindfulness practices such as meditation, yoga, or even simple breathing exercises can help individuals attain a greater sense of inner peace and insight.
- **Transcending Ego and Materialism**: Extraterrestrial encouragement for spiritual growth often involves moving beyond ego-driven pursuits. These messages imply that humans must overcome materialistic and self-centered drives to unlock higher states of consciousness. Extraterrestrial beings are frequently portrayed as having transcended the need for status or material wealth, living instead in service to the collective good. This message challenges individuals to prioritize values such as kindness, service, and humility over material gain or social status.

Collective Spiritual Awakening: Fostering Unity and Connection

Extraterrestrial messages emphasize that individual spiritual growth contributes to a larger, collective awakening. By fostering a sense of unity and interconnectedness, humanity can transcend divisions that often lead to conflict, inequality, and suffering. This call for collective evolution suggests that humanity's future depends on its ability to come together as one, embracing differences as strengths and working collaboratively toward shared goals.

- **Building a Global Community**: The concept of a collective spiritual awakening encourages humans to view themselves as members of a global family. Extraterrestrial messages often advocate for unity, highlighting the importance of transcending ethnic, national, and cultural boundaries. By fostering a sense of global kinship, humanity can lay the foundation for a more peaceful, cooperative world.
- **Creating Spaces for Spiritual Dialogue**: To foster collective growth, humanity can benefit from creating spaces for open dialogue about spirituality and consciousness. Such spaces, whether through educational programs, cultural exchanges, or community gatherings, allow individuals to explore diverse spiritual perspectives and practices. Extraterrestrial messages imply that knowledge-sharing and mutual respect are key to a harmonious society, encouraging individuals to learn from each other's wisdom.
- **Embracing Interconnectedness with All Life**: Many extraterrestrial messages suggest that spiritual growth involves recognizing humanity's interconnectedness with all living beings. This perspective aligns with many Indigenous and Eastern spiritual traditions, which view humanity as an integral part of nature. By respecting and honoring all forms of life, humans can foster a balanced and harmonious relationship with the Earth and each other.

The Higher State of Consciousness: Accessing Wisdom and Intuition

Another recurring theme in extraterrestrial messages is the encouragement to attain higher states of consciousness, often described as realms of expanded perception, wisdom, and intuitive knowledge. These messages suggest that extraterrestrial beings operate from a heightened state of awareness, allowing them to access insights beyond the limitations of ordinary human consciousness. By developing practices that nurture these states, humanity could unlock its own latent potential.

- **Meditative and Mind-Expanding Practices**: Extraterrestrial encouragement often involves meditation or other practices that promote expanded awareness. Through these practices, individuals may experience altered states of consciousness that open pathways to greater insight, creativity, and empathy. Some accounts describe extraterrestrial beings as using these higher states to communicate telepathically or to understand universal truths, implying that humanity could also harness these abilities.
- **Enhancing Intuition and Inner Guidance**: Intuition plays a central role in the development of higher consciousness. Extraterrestrial messages encourage humans to trust their inner guidance, which can serve as a compass for personal growth and decision-making. Practices that foster intuitive awareness, such as journaling, dream analysis, or visualization, can help individuals tune into their inner wisdom and align with their authentic purpose.
- **Accessing Universal Knowledge and Wisdom**: Higher states of consciousness are often portrayed as gateways to universal knowledge, a collective reservoir of wisdom that transcends individual experience. Extraterrestrial beings are sometimes described as drawing on this source of knowledge to impart teachings or guidance to humanity. By cultivating a deeper connection to this reservoir, humans can access insights that promote peace, balance, and harmony.

Spiritual Evolution as Preparation for Cosmic Integration

A significant implication of these messages is that humanity's spiritual evolution is a prerequisite for cosmic integration—a process in which Earth becomes part of a larger interstellar or interdimensional community. Extraterrestrial beings are often described as highly evolved spiritually, embodying values such as compassion, respect, and wisdom. These messages imply that to join this cosmic community, humanity must align itself with these values, preparing to interact responsibly with other civilizations.

- **Demonstrating Readiness for Peaceful Interaction**: The encouragement of spiritual growth suggests that extraterrestrials view spiritual maturity as a marker of readiness for peaceful interstellar relations. Humanity's ability to move beyond war, division, and materialism is seen as a sign of progress, showing that humans are ready to interact as equals with more advanced civilizations. By adopting a mindset of peace, openness, and respect, humanity can signal its readiness for broader interactions.
- **Creating a Harmonious Planetary Culture**: Many messages suggest that humanity must evolve toward a planetary culture rooted in shared values. This culture would transcend current social and political divides, embodying principles of cooperation, stewardship, and compassion. Extraterrestrial beings often describe their own societies as unified by these values, presenting a vision of what humanity could become if it embraces spiritual evolution.

Responding to the Call for Spiritual Growth

The encouragement of spiritual and consciousness evolution in extraterrestrial messages presents humanity with a profound opportunity to elevate its collective mindset. These messages emphasize the importance of compassion, unity, mindfulness, and higher consciousness, urging individuals to embark on personal journeys of inner transformation. By fostering these qualities, humans can contribute to a collective spiritual awakening that transcends boundaries, uniting humanity in a shared pursuit of wisdom, peace, and harmony.

Through practices that nurture mindfulness, empathy, and intuition, individuals can access higher states of consciousness, gaining insights that expand their understanding of self and the universe. As humanity aligns with these values, it not only prepares for peaceful coexistence on Earth but also signals readiness for broader interactions with other civilizations. By heeding this call for spiritual evolution, humanity can transform itself into a compassionate, conscious society capable of contributing positively to a cosmic community, embracing its role as a responsible, enlightened member of a greater interstellar reality.

Cultivating Inner Harmony: Practices and Principles in Alignment with Extraterrestrial Messages on Spiritual Growth

Extraterrestrial messages on humanity's spiritual evolution often highlight the importance of adopting certain practices and principles, such as meditation, nonviolence, and mindfulness, that nurture inner peace and promote harmony. These suggested practices go beyond individual well-being; they contribute to a collective consciousness that resonates with universal values, aligning humanity with higher principles of existence. By integrating practices like meditation and embracing principles such as nonviolence, humanity can transform its inner and outer worlds, becoming more attuned to the needs of the planet, each other, and potentially, the wider cosmic community.

Meditation: Accessing Higher States of Awareness

Meditation stands out as one of the most frequently emphasized practices in extraterrestrial messages focused on spiritual growth. Often described as a pathway to higher consciousness, meditation enables individuals to cultivate inner calm, clarity, and expanded awareness—qualities that many extraterrestrial encounters describe as necessary for humanity's evolution. Reports of extraterrestrial beings often depict them as deeply mindful, suggesting that practices like meditation could bring humans closer to this elevated state of being.

- **The Transformative Power of Meditation**: Meditation helps individuals transcend the everyday thought patterns that often drive anxiety, fear, and division. By fostering inner stillness, meditation enables practitioners to access a state of "higher self," a place of wisdom, compassion, and inner peace. In this state, people can shed personal biases, encouraging an open-minded, interconnected worldview that aligns with the values seen in extraterrestrial messages.
- **Meditation as a Gateway to Expanded Perception**: Many extraterrestrial messages suggest that meditation not only calms the mind but also enhances perception. Meditation enables practitioners to tap into a more intuitive state, allowing them to perceive beyond the limitations of their five senses. This expanded awareness resonates with descriptions of extraterrestrial beings who possess heightened telepathic and empathic abilities. Through meditation, humans can awaken similar latent abilities, deepening their connection to the self and the universe.
- **Integrating Meditation into Daily Life**: To foster sustained spiritual growth, meditation can be integrated into daily routines. Practicing even a few

minutes of meditation each day can build a foundation of mindfulness, equipping individuals to respond thoughtfully to challenges rather than reacting impulsively. Extraterrestrial messages often depict advanced beings as balanced and purposeful, reflecting the mental clarity that meditation cultivates.

Nonviolence: Embracing Compassion as a Way of Life

Extraterrestrial messages often advocate for the principle of nonviolence, emphasizing it as a necessary step in humanity's spiritual evolution. Nonviolence extends beyond physical restraint to include thoughts, words, and actions that reflect compassion and respect for all life forms. This principle aligns with teachings from numerous spiritual traditions, including Buddhism and Jainism, which view nonviolence as essential for creating harmony within oneself and the world.

- **The Principle of Ahimsa**: The concept of nonviolence, or *ahimsa*, invites individuals to avoid causing harm in any form. Ahimsa is more than passive restraint; it is an active choice to foster peace and compassion. Adopting this principle shifts individuals from seeing others as adversaries to perceiving them as fellow beings on a shared journey. This mindset of unity and respect echoes extraterrestrial messages that call for collective empathy and understanding.
- **Cultivating Nonviolence in Thought and Speech**: Nonviolence requires mindfulness in every interaction. Individuals can practice nonviolence by avoiding harmful speech, letting go of resentment, and offering kindness to others—even in challenging situations. Nonviolent communication encourages empathy and honesty, strengthening bonds and reducing misunderstandings. By cultivating this level of mindfulness, humans can embody the harmonious values that extraterrestrial messages highlight.

- **Nonviolence as a Basis for Global Peace**: Many extraterrestrial messages emphasize that humanity must resolve conflicts without resorting to violence if it hopes to join a larger cosmic community. Embracing nonviolence as a cultural standard could lead to more diplomatic approaches to global disputes, setting the foundation for international cooperation. Nonviolence, as practiced at a societal level, could dismantle cycles of retaliation and establish a lasting culture of peace.

Mindfulness: Living in the Present with Purpose

Mindfulness, or the practice of staying fully present, is a foundational practice that supports other spiritual principles. Mindfulness involves paying close attention to one's thoughts, emotions, and surroundings without judgment, creating a space of acceptance and peace. Many extraterrestrial messages suggest that humans can benefit from this practice, as it fosters self-awareness and emotional regulation, which are essential for personal and collective transformation.

- **Developing Presence and Self-Awareness**: Mindfulness enables individuals to observe their thoughts and emotions without identifying with them, creating an inner sense of balance. This practice can reduce impulsivity and encourage a thoughtful response to life's challenges. As many extraterrestrial beings are described as calm and centered, mindfulness can serve as a pathway to emulate this quality of presence in one's daily life.
- **Mindfulness and Emotional Intelligence**: Mindfulness also nurtures emotional intelligence, enabling individuals to manage their emotions and respond compassionately to others. By cultivating emotional awareness, people can avoid projecting their own insecurities or judgments onto others, fostering more authentic relationships. Extraterrestrial messages often suggest that spiritual evolution

involves greater emotional maturity and harmony, both of which mindfulness can promote.
- **Integrating Mindfulness into All Actions**: Practicing mindfulness goes beyond sitting in meditation; it can be applied to all activities, from eating to working to interacting with others. By remaining present in each moment, individuals can bring intention and purpose into their lives. This mindfulness-based approach to life aligns with the extraterrestrial vision of a society where individuals live consciously, grounded in awareness and connection.

Compassion and Service: Building a Heart-Centered Community

Another theme in extraterrestrial messages is the importance of compassion and service to others. Many encounters suggest that highly evolved beings prioritize acts of service, viewing them as expressions of universal love. Practicing compassion and engaging in selfless service can transform individuals and communities, fostering an environment of support, empathy, and interconnectedness.

- **Cultivating Compassion as a Daily Practice**: Compassion invites individuals to place themselves in others' shoes, feeling empathy and seeking to alleviate suffering. Compassion can be cultivated through small, consistent actions—offering a kind word, listening attentively, or extending help when needed. Many extraterrestrial messages reflect this value of compassion, suggesting that humanity can transcend divisions through acts of genuine kindness.
- **Service as a Path to Spiritual Growth**: Engaging in service enables individuals to transcend self-centered concerns and connect with others at a deeper level. Acts of service foster a sense of purpose and fulfillment, reminding people of their interdependence. This approach aligns with extraterrestrial messages that encourage humanity to work together for the

collective good, contributing to a world that values cooperation over competition.
- **Creating a Culture of Compassion**: When individuals embrace compassion and service, they contribute to a ripple effect, inspiring others to adopt similar values. A society that prioritizes compassion and empathy becomes more resilient, supportive, and unified. This culture of compassion aligns with extraterrestrial visions of a harmonious civilization and sets the stage for humanity's advancement.

Integrative Practices: Honoring Interconnectedness Through Ritual and Reflection

In addition to mindfulness, nonviolence, and compassion, many extraterrestrial messages suggest that spiritual practices should honor humanity's interconnectedness with nature, the cosmos, and each other. Rituals that acknowledge this interconnectedness can serve as reminders of humanity's shared journey, helping individuals develop a sense of reverence and responsibility for the greater whole.

- **Rituals of Gratitude and Connection**: Practicing gratitude rituals or spending time in nature can deepen one's appreciation for the interconnected web of life. Such rituals can foster a sense of belonging and responsibility toward the Earth and all beings. By nurturing this sense of connection, individuals become more inclined to act in ways that benefit the planet and each other.
- **Reflective Practices for Self-Alignment**: Reflective practices, such as journaling or silent contemplation, encourage individuals to align with their highest values. By regularly evaluating their thoughts and actions, people can cultivate a life that resonates with the principles highlighted in extraterrestrial messages. Self-reflection allows individuals to course-correct, ensuring that their choices align with values of love, unity, and compassion.

- **Unity with the Cosmos Through Spiritual Symbolism**: Extraterrestrial messages often point toward humanity's place in the cosmic community, inspiring a sense of cosmic belonging. Symbols like the stars, the moon, or the cycles of nature can serve as reminders of humanity's connection to the larger universe. Honoring these symbols in personal rituals or community gatherings can reinforce the spiritual bond between humanity and the cosmos.

Practices and Principles for a New Consciousness

Extraterrestrial messages encouraging spiritual evolution outline specific practices and principles—such as meditation, nonviolence, mindfulness, compassion, and service—that align with humanity's journey toward a higher consciousness. These practices serve as catalysts for personal and collective transformation, fostering inner peace, emotional intelligence, and unity. By adopting these practices, individuals contribute to a global shift in consciousness, one that transcends fear and division, uniting humanity in a shared vision of peace and harmony.

Through meditation, humans can cultivate clarity and intuition, accessing deeper levels of awareness. Nonviolence and compassion invite people to embrace empathy, while mindfulness fosters presence and self-awareness. Service and ritual practices honor the interconnectedness of all life, encouraging humanity to see itself as part of a larger whole. Together, these practices provide a roadmap for humanity to align with the spiritual principles described in extraterrestrial messages, enabling individuals and societies to evolve toward a higher state of being, ready to embrace their place in a universal community.

Summary: Embracing the Urgent Messages for Humanity's Future

The extraterrestrial messages analyzed throughout these topics reveal an interconnected vision for humanity's future, underscored by themes of environmental stewardship, nuclear disarmament, and spiritual growth. The environmental warnings serve as a wake-up call, urging humanity to address climate change, pollution, and biodiversity loss before it's too late. These messages highlight specific actions—conservation efforts, sustainable living, and policies to reduce pollution—that could help restore the Earth's natural balance. By heeding these warnings, humanity not only protects the planet but also honors its responsibility as stewards of life on Earth.

The call for nuclear disarmament speaks directly to the existential threat posed by nuclear weapons. Extraterrestrial messages emphasize that the path to peace and security lies in reducing reliance on nuclear arsenals, inspiring a movement toward global cooperation and disarmament. By recognizing the political and ethical dimensions of these messages, humanity is encouraged to foster diplomatic relations based on trust and mutual respect, building a future that prioritizes dialogue over conflict and deterrence.

Finally, the encouragement for spiritual and consciousness evolution brings a transformative dimension to these messages. Through practices such as meditation, nonviolence, mindfulness, and compassion, individuals are invited to embark on journeys of personal and collective growth. These practices nurture qualities that can reshape societal values, helping humanity transcend materialism, ego, and division. As individuals and communities embrace these principles, they create a harmonious global culture that resonates with the peaceful values embodied by advanced extraterrestrial beings.

Together, these messages offer a roadmap for humanity's future—a vision of an enlightened, peaceful, and interconnected world. By embracing environmental responsibility, fostering peace, and prioritizing spiritual growth, humanity can respond to these messages with purpose and intention, paving the way for a more sustainable and spiritually aware civilization. In doing so, humanity may not only safeguard its future on Earth but also prepare itself for a meaningful place in the larger cosmic community.

Having explored the broader implications of extraterrestrial messages on environmental stewardship, nuclear disarmament, and spiritual growth, it becomes clear that these insights demand meaningful action, both on an individual and collective level. The next logical step is to examine how humanity can actively incorporate these teachings into daily life and societal structures. This involves exploring personal practices—such as mindfulness and sustainable living—that reflect the core themes of these messages, as well as considering ethical lifestyle changes that align with them. Additionally, we will look at how humanity as a whole can mobilize through international initiatives, policies, and social movements to foster a global culture of peace and unity. Finally, we turn to the integration of extraterrestrial messages into the realms of science, philosophy, and religion, which could pave the way for interdisciplinary research and inspire a deeper understanding of humanity's place in the cosmos. By moving from understanding to action, we bridge the gap between these messages and humanity's potential to evolve, both individually and collectively.

Chapter 10: Charting a Path Forward: Personal and Collective Actions for Humanity's Response

The insights offered by extraterrestrial messages call not only for reflection but for action—steps that individuals and societies can take to realign with the values of unity, stewardship, and spiritual growth. This next group of topics delves into how each person can contribute to this vision through mindful practices, ethical lifestyle changes, and a commitment to sustainability. Beyond individual actions, it examines the potential for humanity to organize on a larger scale, implementing international initiatives that support environmental protection and nuclear disarmament, and fostering global unity through social movements. Finally, it addresses how integrating these messages into scientific, philosophical, and religious discourse can enrich humanity's understanding of its role in the universe. By blending personal dedication with collective efforts and interdisciplinary inquiry, humanity can embody the transformative potential embedded in these messages, creating a future that aligns with both planetary and cosmic harmony.

Embracing Personal Transformation: Practices to Align with Extraterrestrial Messages

In the face of extraterrestrial messages that emphasize environmental responsibility, spiritual evolution, and unity, individuals are uniquely positioned to make meaningful changes in their daily lives. While global policies and social movements have their role, the journey toward a more harmonious world begins with each person's choices and actions. Personal practices, such as mindfulness and sustainability, empower individuals to resonate with the values embedded in these messages and contribute to the broader goal of planetary harmony. By fostering mindful awareness, individuals can cultivate a deeper connection to themselves, others, and the environment, embodying the principles of compassion and responsibility that many extraterrestrial messages encourage.

Mindfulness: Cultivating Inner Awareness and Balance

Mindfulness, the practice of being fully present in the moment without judgment, has become a cornerstone of spiritual and psychological well-being. By cultivating mindfulness, individuals can develop greater self-awareness, emotional intelligence, and a sense of inner peace, aligning themselves with the values of higher consciousness that extraterrestrial messages often promote. Mindfulness serves as both a tool for personal growth and a foundation for positive interactions with others and the environment.

- **The Power of Presence**: Mindfulness encourages individuals to observe their thoughts, feelings, and surroundings without becoming overwhelmed by them. In a world filled with distractions and constant stimuli, this practice brings clarity and focus, helping people respond thoughtfully rather than react impulsively. Stories of extraterrestrial encounters often describe beings who exude calm and presence, suggesting that such qualities are hallmarks of spiritual maturity. By practicing mindfulness, individuals can develop similar qualities, which can foster harmony in their personal lives and interactions.
- **Mindfulness and Compassion**: Through mindfulness, individuals can cultivate empathy by observing their own emotions and recognizing similar feelings in others. Compassion naturally arises when one becomes attuned to their shared human experience,

dissolving barriers between the self and others. Extraterrestrial messages often emphasize compassion as a central theme, suggesting that humanity's spiritual evolution depends on the ability to feel and act with kindness. Mindfulness fosters this compassionate approach to life, promoting unity and understanding in personal and collective interactions.

- **Daily Mindfulness Practices**: Incorporating mindfulness into daily routines doesn't require significant time or effort. Simple practices, such as mindful breathing, taking moments of pause, or observing nature with intention, can enhance one's awareness and presence. These small acts of mindfulness build over time, transforming ordinary moments into opportunities for spiritual growth. By fostering an appreciation for the present moment, individuals can experience life more fully, embracing a harmonious existence that aligns with the extraterrestrial messages encouraging spiritual awakening.

Sustainability: Practicing Environmental Stewardship

The call for sustainability is one of the most consistent themes in extraterrestrial messages, highlighting the urgency of preserving Earth's natural resources and ecosystems. Embracing sustainable practices allows individuals to embody their role as caretakers of the planet, promoting a lifestyle that respects the interconnectedness of all life. Sustainability is not only an environmental practice; it is a moral commitment to live in a way that respects both current and future generations.

- **Reducing Consumption**: One of the simplest and most impactful ways individuals can embrace sustainability is by reducing unnecessary consumption. This can include mindful shopping, choosing quality over quantity, and avoiding single-use items. By consuming less, individuals reduce their ecological footprint, preserving resources for future generations. Extraterrestrial messages often warn against the consequences of materialism and resource depletion, suggesting that humanity's survival depends on adopting a more balanced approach to consumption.

- **Supporting Eco-Friendly Products and Companies**: Choosing sustainable products and supporting environmentally responsible companies helps individuals make a positive impact beyond their immediate lives. When individuals prioritize eco-friendly alternatives—such as organic produce, sustainable clothing, and renewable energy—they send a message to the marketplace about the importance of environmental responsibility. This shift in consumer behavior reflects the extraterrestrial messages urging humanity to honor the planet and adopt a long-term perspective on resource management.
- **Practicing Minimalism and Mindful Ownership**: Minimalism, or the practice of owning only what is necessary and meaningful, encourages individuals to let go of material excess. By focusing on what truly adds value to their lives, people can create more space for experiences, relationships, and personal growth. Minimalism aligns with the extraterrestrial messages that emphasize spiritual values over material possessions, offering a pathway to personal fulfillment that does not rely on external wealth or status.

Embracing Renewable Energy and Reducing Carbon Footprint

Adopting practices that reduce one's carbon footprint plays a significant role in mitigating climate change and preserving the environment. By making small adjustments, such as using renewable energy sources, individuals can contribute to the reduction of greenhouse gas emissions, reflecting a commitment to environmental responsibility.

- **Switching to Renewable Energy**: Choosing renewable energy sources, like solar or wind power, aligns with the extraterrestrial messages that encourage environmental stewardship. By reducing reliance on fossil fuels, individuals can lessen their environmental impact and support the transition to a sustainable energy system. Many utility companies offer renewable energy options, allowing individuals to choose greener alternatives for their household energy needs.
- **Reducing Vehicle Emissions**: Transportation is a significant source of pollution, and small changes in daily habits can make a considerable difference. Opting for public transit, carpooling, cycling, or walking reduces individual carbon footprints and supports a cleaner environment. In the context of extraterrestrial messages, these actions embody the idea of living harmoniously with the planet by making conscious choices that respect the environment.
- **Energy Efficiency at Home**: Practicing energy efficiency—such as using energy-efficient appliances, reducing electricity use, and insulating homes properly—helps conserve resources. These actions not only reduce household energy costs but also contribute to larger environmental efforts. Energy efficiency aligns with the extraterrestrial messages calling for responsible resource use, as it reflects a commitment to preserving Earth's limited resources for future generations.

Fostering a Mindful Connection with Nature

Many extraterrestrial messages encourage humanity to reconnect with nature, suggesting that this bond is crucial for understanding and respecting the planet. Spending time in natural settings can inspire gratitude and foster an appreciation for the delicate ecosystems that sustain life. By cultivating this connection, individuals can experience a sense of unity with the Earth, leading to more environmentally conscious decisions.

- **Engaging in Outdoor Activities Mindfully**: Whether hiking, gardening, or simply observing wildlife, spending time outdoors can deepen one's relationship with nature. Practicing mindfulness during these activities—such as listening to the sounds of birds, feeling the texture of plants, or observing natural patterns—reinforces a sense of interconnectedness with the environment. This connection aligns with extraterrestrial messages that highlight the importance of unity and balance within Earth's ecosystems.
- **Nature-Inspired Reflection and Gratitude**: Reflecting on the beauty and diversity of nature can inspire gratitude for Earth's resources. Practicing gratitude strengthens one's resolve to protect the environment, transforming sustainable practices into acts of reverence for life. Such gratitude-based practices resonate with the extraterrestrial messages calling for humanity to recognize its role as a steward of Earth, fostering a mindset of respect and humility.
- **Participating in Environmental Restoration Efforts**: Volunteering for tree-planting initiatives, beach clean-ups, or other environmental restoration projects allows individuals to actively contribute to the health of their communities and ecosystems. Extraterrestrial messages often warn of the dangers of environmental degradation, and by engaging in restoration work, individuals take tangible steps toward reversing these effects.

The Power of Personal Action in Shaping Humanity's Future

Extraterrestrial messages encourage humanity to embody values of mindfulness, sustainability, and environmental stewardship, suggesting that the path to a harmonious future lies within each person's actions. Through daily practices of mindfulness, individuals develop greater self-awareness, empathy, and clarity, aligning themselves with the calm, centered qualities observed in evolved beings. Sustainability practices allow people to embody the role of Earth's caretakers, making choices that reflect respect for the planet's resources and ecosystems.

By integrating these practices into their lives, individuals contribute to a collective transformation, fostering a society that values spiritual growth, environmental responsibility, and interdependence. As each person embraces these values, humanity moves closer to the vision outlined in extraterrestrial messages—a world where peace, unity, and ecological balance form the foundation of human existence. Personal action, though small in scale, can ripple outward, inspiring others and shaping a future where humanity lives in harmony with the planet and the universe.

Aligning with Higher Values: Ethical Principles and Lifestyle Changes for Humanity's Future

The call to action in extraterrestrial messages encourages individuals to live ethically and mindfully, embodying values that transcend personal gain and prioritize collective well-being. By embracing ethical principles and making lifestyle changes that reflect these values, individuals contribute to a more harmonious and sustainable future. This shift involves not only adopting values like kindness, respect, and empathy but also reassessing one's relationship with material possessions, environmental resources, and social responsibilities. Embracing these principles creates a ripple effect, inspiring communities and, eventually, the world to evolve toward a more conscious and interconnected existence.

Cultivating Compassion and Empathy in Daily Life

Compassion and empathy form the bedrock of an ethical lifestyle that aligns with the messages often conveyed in extraterrestrial encounters. These values invite individuals to consider the feelings, needs, and perspectives of others, whether in personal relationships or broader social interactions. By cultivating compassion, individuals can foster harmonious relationships and contribute to a world that values understanding over conflict.

- **Empathy as a Mindful Practice**: Empathy involves actively listening to others without judgment, allowing individuals to see beyond differences and recognize shared humanity. Practicing empathy transforms interactions from self-centered exchanges into opportunities for connection and growth. Many extraterrestrial messages encourage humanity to embrace empathy as a means of bridging divides and fostering peace.
- **Compassionate Actions as a Daily Commitment**: Compassion goes beyond feelings of kindness—it manifests through actions that alleviate suffering. Small acts of kindness, such as helping a neighbor or volunteering, embody compassion and create a positive impact on those around us. Stories from various spiritual traditions and extraterrestrial messages suggest that compassion is an essential attribute for personal and collective evolution, highlighting its power to unite and heal communities.

Integrity and Honesty: Living with Authenticity and Transparency

Integrity and honesty are cornerstones of an ethical life, fostering trust and authenticity. These values guide individuals to act in alignment with their inner truths, regardless of external pressures. Honesty nurtures transparent relationships, creating a foundation for mutual respect and understanding. Living with integrity involves honoring commitments, being truthful, and standing up for ethical principles—even in challenging situations.

- **Honesty in Self-Reflection and Actions**: Integrity begins with self-honesty, where individuals assess their values, beliefs, and motivations. This introspection allows individuals to understand their authentic selves and to make choices that align with their inner principles. Many extraterrestrial messages advocate for clarity of intention and honesty as essential qualities for spiritual growth, suggesting that integrity contributes to inner peace and societal stability.
- **Transparency in Relationships and Communication**: Honesty fosters genuine connections and encourages open communication. By practicing transparency in personal and professional relationships, individuals promote an environment of trust and mutual respect. Extraterrestrial messages often portray advanced civilizations as transparent and trustworthy, suggesting that humanity can achieve harmony by embracing similar principles in its interactions.

Simplicity and Minimalism: Reassessing Material Needs

Extraterrestrial messages frequently warn against the excesses of materialism, urging humanity to prioritize spiritual values over material possessions. Adopting simplicity and minimalism as lifestyle choices encourages individuals to reduce their reliance on material goods, focusing instead on experiences, relationships, and personal growth. Minimalism is not about deprivation; rather, it is a conscious choice to own and consume only what is necessary and meaningful.

- **Mindful Consumption and Resource Management**: Choosing to live simply involves mindful consumption, where individuals consider the environmental and ethical implications of their purchases. By opting for quality over quantity, individuals reduce waste, conserve resources, and support sustainable production practices. Extraterrestrial messages highlight the importance of preserving the Earth's resources, suggesting that simplicity can lead to a more balanced relationship with the environment.
- **Embracing Minimalism in Daily Life**: Minimalism encourages individuals to focus on what truly matters, eliminating distractions that impede inner peace and spiritual growth. By owning fewer material possessions, individuals have more time and energy to devote to relationships, personal development, and contributing to the well-being of others. This lifestyle reflects the values embedded in extraterrestrial messages, where spiritual fulfillment takes precedence over material gain.

Environmental Responsibility: Making Choices that Reflect Earth Stewardship

Many extraterrestrial messages emphasize the importance of environmental responsibility, urging humanity to adopt lifestyles that respect the Earth and its ecosystems. Embracing environmental ethics involves making choices that reduce one's ecological footprint, from conserving energy to supporting eco-friendly products. By practicing environmental stewardship, individuals honor their role as caretakers of the planet, contributing to a sustainable future.

- **Reducing Waste and Embracing Recycling**: Reducing waste is a simple yet impactful way individuals can minimize their environmental impact. This can include recycling materials, reducing plastic use, and avoiding disposable items. Extraterrestrial messages often warn of the consequences of environmental neglect,

suggesting that humanity's survival depends on embracing sustainable practices that prevent further ecological damage.
- **Supporting Sustainable and Local Food Systems**: Choosing organic and locally sourced foods helps reduce the carbon footprint associated with transportation and industrial agriculture. It also supports local farmers who practice environmentally responsible farming. By opting for sustainable food choices, individuals embody the extraterrestrial message that encourages environmental consciousness, contributing to healthier ecosystems and communities.

Nonviolence and Ethical Treatment of All Beings

Nonviolence extends beyond human relationships to encompass a respect for all forms of life. This principle aligns with extraterrestrial messages that often emphasize unity, compassion, and harmony. Practicing nonviolence involves rejecting harmful practices toward animals and nature, embracing instead a lifestyle of respect and reverence for all living beings. For many, this value may inspire dietary choices, such as adopting a vegetarian or vegan diet, or supporting ethical practices in industries that impact animals.

- **Compassionate Diet Choices**: Choosing plant-based or ethically sourced foods reflects a commitment to nonviolence and respect for life. By opting for diets that reduce harm to animals and ecosystems, individuals embrace the extraterrestrial message of compassion and interconnectedness, contributing to a world that values ethical treatment of all beings.
- **Nonviolence in Daily Interactions**: Nonviolence also applies to human relationships, where individuals commit to resolving conflicts peacefully, avoiding harmful language, and practicing patience. By embracing nonviolence in thoughts, words, and

actions, individuals promote peace within themselves and their communities. This principle is a common theme in extraterrestrial messages, which depict evolved beings as inherently peaceful, suggesting that humanity's growth requires a commitment to nonviolence.

Commitment to Lifelong Learning and Self-Improvement

Extraterrestrial messages often encourage humanity to strive for personal and spiritual growth, implying that the journey toward self-improvement is essential for individual and collective evolution. Committing to lifelong learning and self-reflection enables individuals to cultivate qualities like humility, resilience, and wisdom, which contribute to a balanced and fulfilling life. This commitment also supports open-mindedness and adaptability, qualities that prepare individuals for potential future interactions with other intelligent beings.

- **Pursuing Knowledge and Wisdom**: Embracing lifelong learning fosters curiosity and a willingness to question assumptions. By seeking knowledge through education, travel, and meaningful experiences, individuals broaden their perspectives and develop empathy for others. Extraterrestrial messages often portray advanced civilizations as knowledgeable and wise, suggesting that humanity can prepare for future growth by prioritizing continuous learning.
- **Practicing Self-Reflection and Growth**: Self-reflection allows individuals to assess their strengths, weaknesses, and motivations, guiding them toward personal growth. By regularly evaluating their thoughts, behaviors, and relationships, individuals can make adjustments that bring them closer to their highest ideals. This process of self-improvement aligns with extraterrestrial messages that emphasize the

importance of introspection and self-awareness in spiritual evolution.

Living a Life Aligned with Higher Ethical Principles

The ethical principles highlighted in extraterrestrial messages call on individuals to embody values of compassion, integrity, simplicity, environmental responsibility, nonviolence, and lifelong learning. These values serve as guiding lights, shaping a lifestyle that resonates with spiritual growth, collective well-being, and a profound respect for all life. By practicing empathy, individuals nurture connections based on mutual respect. Living with integrity and honesty creates authentic relationships and fosters trust, while embracing simplicity and environmental responsibility helps humanity preserve its precious resources.

Adopting a lifestyle of nonviolence and ethical treatment toward all beings reflects a commitment to peace and compassion, resonating with the harmonious qualities described in extraterrestrial messages. Finally, committing to lifelong learning and self-reflection enables individuals to grow continuously, embracing the humility and wisdom essential for personal and collective evolution.

By integrating these ethical principles into their daily lives, individuals can embody the values that extraterrestrial messages seek to inspire. This alignment with higher values has the potential to transform not only individual lives but also the broader society, contributing to a world that honors unity, respect, and interconnectedness with all life. Through these choices, humanity can create a future that reflects the ideals of an enlightened civilization, prepared for harmonious existence on Earth and within the greater cosmic community.

A Unified Front: International Environmental Initiatives for Global Harmony

The messages received in various extraterrestrial encounters often emphasize the importance of environmental stewardship, urging humanity to take urgent, cooperative actions to protect and restore the Earth. Addressing environmental challenges at a global scale requires more than individual lifestyle changes; it demands concerted efforts through international policies and initiatives. By creating binding treaties and cooperative frameworks, countries worldwide can align toward the shared goal of safeguarding the planet for future generations. Understanding the potential for international environmental initiatives means examining the frameworks, treaties, and collaboration strategies that can help humanity address ecological crises, such as climate change, biodiversity loss, and resource depletion.

Setting the Foundation: Historical Context of Environmental Agreements

The history of international environmental agreements offers valuable insight into how global cooperation can drive effective action on environmental issues. Treaties like the 1992 Rio Earth Summit, the Kyoto Protocol, and the Paris Agreement represent major milestones in humanity's collective response to environmental threats. By studying these agreements, we can gain a clearer picture of the obstacles, successes, and lessons learned in the quest for environmental sustainability.

- **The Rio Earth Summit**: The 1992 Rio Earth Summit, or United Nations Conference on Environment and Development (UNCED), brought global attention to sustainability. Nations agreed on the principles of

sustainable development, recognizing the need for a balance between economic growth, environmental preservation, and social equity. The summit also produced the landmark Agenda 21, a non-binding action plan that highlighted humanity's responsibility to act as stewards of the Earth, reflecting themes found in extraterrestrial messages.
- **The Kyoto Protocol**: Established in 1997, the Kyoto Protocol was the first binding international agreement focused on reducing greenhouse gas emissions. It set targets for developed countries, marking a global commitment to mitigating climate change. While the protocol faced challenges in implementation, it underscored the necessity of binding commitments in achieving substantial environmental progress.
- **The Paris Agreement**: Adopted in 2015, the Paris Agreement marked a significant step toward global climate action by committing countries to limit global temperature rise to well below 2°C above pre-industrial levels. Unlike previous agreements, the Paris Agreement emphasized a collaborative, country-driven approach, where each nation set its own emissions targets. This flexible structure has allowed for widespread participation and shows the importance of adaptable frameworks in fostering global cooperation.

Creating Future Pathways: The Need for New International Environmental Initiatives

While existing agreements have made strides, the challenges of climate change, deforestation, and biodiversity loss demand additional, innovative approaches. Building on the successes and lessons of past initiatives, future international environmental efforts must be ambitious, inclusive, and resilient to address the complexities of the modern world.

- **Addressing Global Deforestation**: Forests play a critical role in absorbing carbon dioxide and supporting biodiversity. International initiatives focusing on protecting forests—such as REDD+ (Reducing Emissions from Deforestation and Forest Degradation)—have helped address deforestation, but a comprehensive global framework is needed to ensure protection for all major forests. Such an initiative could involve financial incentives, enforcement mechanisms, and local community engagement to preserve forest ecosystems.

- **Mitigating Biodiversity Loss**: The Convention on Biological Diversity (CBD) serves as the primary framework for global biodiversity conservation, but its goals often lack enforceable mechanisms. Future treaties could include more binding commitments to preserve biodiversity hotspots and protect endangered species. International cooperation could also promote habitat restoration projects and rewilding efforts, helping to recover ecosystems that have been damaged by human activity.
- **Reducing Plastic Pollution and Ocean Protection**: Plastic pollution is a growing global crisis, affecting marine life and contaminating ecosystems. The United Nations has initiated discussions around a legally binding global treaty to combat plastic pollution, which would include reducing plastic production, increasing recycling, and eliminating single-use plastics. Protecting oceans through coordinated efforts, such as marine protected areas (MPAs) and sustainable fishing regulations, can prevent further damage to marine ecosystems and maintain their biodiversity.

The Role of Climate Finance: Supporting Developing Nations

For international environmental agreements to be successful, they must address the economic disparities between developed and developing nations. Climate finance, which involves transferring funds from wealthy countries to support sustainable projects in less affluent countries, is crucial for achieving equitable environmental solutions. Without adequate funding and resources, developing countries may struggle to meet environmental goals while addressing economic needs.

- **The Green Climate Fund (GCF)**: Established under the Paris Agreement, the Green Climate Fund supports climate adaptation and mitigation in developing countries. By providing financial resources, the GCF empowers nations to pursue renewable energy, reforestation, and sustainable agriculture projects. This form of support is essential in bridging the gap between environmental commitments and economic realities, reflecting the extraterrestrial message of unity and shared responsibility for the planet.
- **Technology Transfer**: Alongside financial aid, developed countries can share technology and expertise with developing nations, enabling them to adopt cleaner energy sources and sustainable practices. Renewable energy technologies, such as solar and wind power, offer cleaner, cost-effective solutions to fossil fuels, but access to these technologies remains limited in some regions. Facilitating technology transfer can help create a level playing field, allowing all countries to contribute to environmental goals.

Building a Global Culture of Environmental Responsibility

In addition to formal treaties, the global community can foster environmental consciousness through education, awareness campaigns, and cross-cultural exchange. Cultivating a mindset of stewardship and interconnectedness encourages individuals, communities, and nations to act in harmony with the Earth. By promoting environmental education and awareness, the international community can inspire future generations to prioritize sustainability, unity, and empathy for the planet.

- **Environmental Education and Youth Empowerment**: Education initiatives that teach sustainability principles and environmental responsibility can help cultivate a generation of environmentally conscious leaders. Programs that engage youth in conservation, waste reduction, and climate action foster a sense of responsibility, encouraging them to champion environmental causes within their communities and beyond.
- **Cross-Cultural Environmental Initiatives**: Cross-cultural initiatives, such as international youth exchanges, conservation projects, and research collaborations, can bridge cultural divides and unite people around shared environmental goals. By collaborating on conservation efforts, people from different backgrounds can build mutual respect and a shared understanding of the planet's significance, aligning with the extraterrestrial messages of unity.
- **International Awareness Campaigns**: Global campaigns, such as Earth Day and World Environment Day, raise awareness about environmental issues and inspire action at the grassroots level. Campaigns that highlight specific issues, like plastic pollution or deforestation, mobilize people across the globe, emphasizing that environmental responsibility transcends national boundaries.

Embracing the Extraterrestrial Call for Unity and Environmental Stewardship

The emphasis on environmental protection in extraterrestrial messages highlights the need for a collective, global approach to planetary challenges. By embracing international treaties and initiatives, humanity acknowledges its responsibility as Earth's guardian and takes steps toward a sustainable future. Global cooperation not only amplifies environmental impact but also aligns with the extraterrestrial vision of unity, where nations work together to ensure a healthy and balanced Earth for all.

Summary: The Imperative of Global Environmental Collaboration

Addressing the environmental challenges of today requires more than isolated actions; it demands a unified, global response. International environmental initiatives and treaties represent humanity's commitment to overcoming the crises that threaten ecosystems, climate stability, and biodiversity. From the historical groundwork laid by agreements like the Rio Earth Summit and the Paris Agreement to emerging efforts in forest conservation, biodiversity protection, and ocean preservation, international cooperation continues to evolve as a powerful tool for sustainable change.

Through climate finance, technology transfer, and educational programs, countries can support one another in reaching shared environmental goals. By fostering a culture of environmental responsibility, humanity can cultivate a global consciousness that values sustainability, interdependence, and empathy for all life on Earth. This alignment with extraterrestrial messages underscores humanity's potential to evolve as a cooperative, enlightened species capable of preserving the planet for future generations.

Mobilizing for a Unified World: The Role of Social Movements in Advancing Peace and Global Unity

Throughout history, social movements have been powerful catalysts for transformation, uniting people under shared values and creating change that ripples through societies. Today, as extraterrestrial messages increasingly encourage humanity to embrace unity, peace, and global interconnectedness, social movements geared toward these ideals hold unparalleled potential to reshape our world.

These movements transcend national borders, uniting individuals from diverse backgrounds in a shared commitment to a harmonious, sustainable, and peaceful future. By analyzing the structure, strategies, and impact of global peace movements, we gain insight into how these collective efforts can align humanity with the values conveyed in extraterrestrial messages and foster a world rooted in unity and understanding.

Uniting Humanity: The Emergence of Global Peace Movements

Global peace movements often arise from humanity's shared desire to end violence, confront injustice, and address systemic inequalities. As they mobilize people across different regions and cultures, these movements emphasize humanity's interconnectedness and the importance of peaceful coexistence. From the anti-war demonstrations of the 1960s to contemporary initiatives addressing climate justice and humanitarian crises, global peace movements have historically served as touchstones for the collective desire to transcend conflict and foster harmony.

- **The Anti-War Movements of the 1960s and 1970s**: The Vietnam War catalyzed one of the most significant global peace movements in modern history, inspiring millions worldwide to protest against military aggression and advocate for diplomatic resolutions. This movement created a legacy of pacifism that persists today, setting the stage for future peace-oriented activism. Extraterrestrial messages often warn against violence and conflict, resonating with the spirit of these movements by urging humanity to embrace peace over war.
- **The Global Climate Movement**: While primarily environmental, the global climate movement also represents a call for unity and shared responsibility.

The effects of climate change transcend borders, reminding humanity of its interconnectedness and the need for collective action. Movements like Fridays for Future, initiated by youth activist Greta Thunberg, have highlighted not only environmental stewardship but also the social responsibility to protect future generations. The movement's emphasis on shared planetary responsibility echoes extraterrestrial messages that underscore the importance of caring for Earth as a unified civilization.

- **The Modern Social Justice Movement**: Social justice movements focused on equality, human rights, and systemic reform highlight humanity's pursuit of fairness and dignity for all. Movements like Black Lives Matter, #MeToo, and indigenous rights activism demonstrate a collective resistance to oppression and advocate for a world where peace and unity are achieved through justice and mutual respect. These movements align with extraterrestrial messages that encourage compassion, inclusivity, and the recognition of each individual's intrinsic value.

Building Momentum: Strategies for Lasting Impact in Peace Movements

For social movements to bring about meaningful change, they must employ strategic frameworks that effectively mobilize people, create awareness, and advocate for policy reform. The most successful movements are those that balance grassroots activism with institutional engagement, building a broad coalition that spans from local communities to international organizations.

- **Grassroots Organizing and Community Engagement**: Effective peace movements often begin at the grassroots level, mobilizing communities through awareness campaigns, public demonstrations, and educational initiatives. By connecting with individuals on a personal level, grassroots movements

foster a sense of collective responsibility and empower people to take part in the mission. Community-based peace efforts align with the extraterrestrial message of unity, demonstrating that change begins within each individual and spreads outward.
- **Leveraging Media and Technology**: In the digital age, social media and technology have revolutionized the ways in which movements spread their message and gather support. Online platforms provide global visibility, enabling activists to share stories, mobilize protests, and coordinate actions with unprecedented speed. The viral nature of social media amplifies the impact of peace movements, making it easier to raise awareness about pressing issues and invite participation from a worldwide audience. As extraterrestrial messages emphasize unity, the interconnectedness fostered by digital platforms can serve as a means of embodying this message in action.
- **Policy Advocacy and Institutional Change**: To achieve lasting change, peace movements often seek to influence policy through advocacy, lobbying, and engagement with governmental and international organizations. By translating grassroots momentum into actionable demands, movements can enact systemic reforms that promote peace and unity on a large scale. The United Nations, for example, often collaborates with global movements to address peace, security, and humanitarian concerns, reflecting an institutional recognition of the values extraterrestrial messages urge humanity to embrace.

Educating for Peace: The Power of Awareness and Knowledge

Education is a vital component of any movement that seeks to create long-term cultural and social transformation. Peace education initiatives teach individuals to value conflict resolution, empathy, and global citizenship, laying the groundwork for a future where unity and cooperation take precedence over division and hostility.

- **Promoting Conflict Resolution Skills**: Education that emphasizes nonviolent communication, mediation, and conflict resolution prepares individuals to handle disagreements constructively. By instilling these skills in young people, peace education fosters a culture of understanding and respect, essential for maintaining harmony on a global scale. Extraterrestrial messages that advocate for peace often suggest that humanity must develop the tools for resolving conflicts amicably, beginning with individual skills and extending to international relations.
- **Cultivating Empathy and Cultural Awareness**: Peace education often includes curricula that explore cultural diversity, empathy, and global interdependence, teaching students to appreciate differences while recognizing shared human values. By nurturing empathy, education enables future generations to act compassionately toward others and transcend cultural barriers, resonating with the extraterrestrial encouragement of a unified, compassionate world.
- **Global Citizenship and Shared Responsibility**: Peace education encourages individuals to think of themselves as global citizens, emphasizing shared responsibility for the planet's well-being. This perspective fosters a collective consciousness that prioritizes humanity's future over individual interests. The concept of global citizenship directly aligns with extraterrestrial messages that emphasize interconnectedness and the need for humanity to act as stewards of the Earth.

Fostering a Culture of Nonviolence: Integrating Peaceful Ideals into Society

A culture of nonviolence requires a collective commitment to resolving conflicts without aggression, both within communities and between nations. By embracing nonviolence, humanity can move toward a world where disagreements are addressed through dialogue, respect, and mutual understanding.

- **Nonviolence in International Relations**: Diplomatic efforts, peace treaties, and conflict prevention programs aim to resolve international disputes without resorting to warfare. Organizations like the United Nations and peacekeeping NGOs play a vital role in mediating conflicts and promoting diplomacy. Nonviolence in international relations reflects the extraterrestrial guidance for harmony, signaling humanity's willingness to transcend destructive behaviors in favor of unity.
- **Community-Based Nonviolence Programs**: Local initiatives that promote nonviolence—such as anti-bullying campaigns, restorative justice practices, and community dialogue forums—can serve as microcosms of global peace efforts. These programs foster social cohesion and reduce conflict, showing that peace begins within communities before extending outward. Extraterrestrial messages often emphasize the importance of local action as a foundation for global change, aligning with community-based nonviolence efforts.
- **Modeling Peaceful Leadership**: Leaders who exemplify peaceful, compassionate, and nonviolent values set a powerful example for society. Leaders such as Mahatma Gandhi, Martin Luther King Jr., and the Dalai Lama have demonstrated the transformative power of nonviolence, inspiring generations to pursue justice through peaceful means. The extraterrestrial message of unity and peace resonates strongly with the

principles of such leaders, who show that strength and influence need not come from force but from a commitment to compassion and integrity.

The Path to Global Unity through Social Movements

Social movements that advocate for peace and global unity serve as powerful reflections of the values promoted in extraterrestrial messages. By mobilizing individuals around shared ideals, these movements foster a culture of understanding, nonviolence, and empathy, providing humanity with a blueprint for a harmonious future. Whether addressing environmental justice, human rights, or social equity, peace movements demonstrate the strength of collective action and the potential for global change.

Through grassroots organizing, educational initiatives, and policy advocacy, these movements inspire individuals to transcend personal and national interests in favor of a collective vision for peace. As humanity continues to respond to calls for unity, social movements will play an essential role in guiding individuals and communities toward a future that aligns with the extraterrestrial vision of a united, compassionate, and peaceful world. In embracing these movements, humanity steps closer to fulfilling its potential as stewards of Earth, ready to engage with the universe in harmony.

Bridging Worlds: The Role of Science, Philosophy, and Religion in Understanding Extraterrestrial Messages

The integration of extraterrestrial messages into human thought systems—science, philosophy, and religion—represents an extraordinary opportunity for expanding humanity's perspective on life, consciousness, and responsibility. Each of these fields brings a unique lens through which to explore these messages, contributing valuable insights into their meaning and implications for the future. By harmonizing scientific inquiry, philosophical analysis, and religious wisdom, humanity can build a more complete understanding of extraterrestrial communications, fostering a more unified worldview that aligns with the messages' call for peace, unity, and environmental stewardship.

Science and Extraterrestrial Messages: A Rational Framework for Exploration

In the realm of science, extraterrestrial messages are often approached with skepticism but also with a keen interest in understanding their origins, nature, and credibility. Scientists and researchers use empirical methods to assess reports of extraterrestrial encounters, evaluate the authenticity of evidence, and explore the potential implications for fields such as astronomy, biology, and psychology. As science is rooted in evidence and systematic investigation, it provides a framework for rigorously examining extraterrestrial messages and their impact on human thought.

Astrobiology and the Search for Life

Astrobiology, the study of life beyond Earth, stands at the forefront of scientific interest in extraterrestrial messages. By exploring the possibility of life on other planets, astrobiologists attempt to answer fundamental questions about the conditions required for life and the likelihood of intelligent civilizations elsewhere in the universe.

- **Understanding Alien Life Forms**: Messages purportedly from extraterrestrial beings often describe physical characteristics, environmental preferences, and modes of communication that are vastly different from those on Earth. Astrobiologists analyze these descriptions to speculate on the biological adaptations and ecological factors that might sustain such life forms, grounding extraterrestrial messages within the context of known biological principles.

- **Exoplanetary Research and Habitability**: The discovery of exoplanets—planets orbiting other stars—has opened new possibilities for locating life beyond our solar system. Research in this area examines factors like atmospheric composition, water presence, and temperature to assess whether a planet could support life. Scientific inquiry into exoplanet habitability resonates with extraterrestrial messages suggesting that humanity is part of a larger cosmic community.

Neuroscience and the Mind's Role in Perception

Neuroscience provides valuable insights into how humans perceive and process experiences, including encounters that involve telepathic or visual communication with extraterrestrial beings. By examining brain function, memory, and altered states of consciousness, neuroscientists seek to understand the mechanisms underlying experiences of contact, whether they stem from external sources or internal cognitive processes.

- **Exploring Telepathic Communication**: Some messages from extraterrestrial sources are reported to occur telepathically. Neuroscience research on brainwave patterns, neural networks, and consciousness attempts to explore the possibility of direct thought transfer, contributing a rational foundation to this otherwise mysterious form of communication.
- **Memory and Reliability**: Encounters with extraterrestrial beings are often remembered vividly,

but the accuracy of these memories can be questioned. Neuroscientific research examines memory formation and recall, assessing the impact of trauma, stress, and altered states on an individual's perception of events. Understanding memory reliability helps distinguish between objective experiences and subjective interpretations, adding clarity to reports of extraterrestrial messages.

Quantum Physics and Interdimensional Theory

Quantum physics has introduced groundbreaking concepts that challenge conventional understandings of reality, such as the potential for multiple dimensions and entanglement. Some theories propose that extraterrestrial beings may exist in dimensions beyond the visible spectrum or communicate through quantum processes, offering a scientific basis for phenomena previously relegated to the realm of mysticism.

- **Interdimensional Communication**: Quantum theories that suggest alternate dimensions or parallel realities open the possibility of communication with beings that inhabit these dimensions. Such hypotheses resonate with reports of extraterrestrial beings appearing momentarily or interacting in ways that defy physical laws, suggesting that they may exist within dimensions beyond human perception.
- **Entanglement and Non-Locality**: Quantum entanglement, where particles remain connected across vast distances, could provide a basis for understanding how extraterrestrial messages are transmitted instantaneously. By studying entanglement, scientists may one day unlock mechanisms that explain telepathic messages and instantaneous communication across light-years.

Philosophy and Extraterrestrial Messages: Expanding Humanity's Ethical Horizons

Philosophy offers a profound space for exploring the implications of extraterrestrial messages, encouraging humans to reflect on questions of existence, morality, and purpose. The philosophical approach to extraterrestrial messages allows humanity to ponder its place in the universe, the nature of intelligence and consciousness, and the ethical responsibilities that may arise from interstellar contact.

The Nature of Consciousness and Intelligence

Extraterrestrial messages challenge humanity to redefine traditional concepts of intelligence and consciousness. Philosophical inquiries into these topics ask questions about what it means to be a sentient being, how intelligence should be measured, and whether consciousness is an inherent quality that extends beyond Earth.

- **Redefining Intelligence**: Traditional human concepts of intelligence are often anthropocentric. Encounters with extraterrestrial intelligence prompt philosophers to reconsider intelligence as a multi-dimensional quality that may manifest differently across species. This redefinition encourages humans to approach extraterrestrial messages with openness, avoiding biases that stem from human-centric assumptions.
- **Consciousness and the Universe**: Some philosophical theories propose that consciousness is a fundamental property of the universe, present in all beings to varying degrees. This idea aligns with extraterrestrial messages that suggest a shared cosmic awareness, encouraging humans to see themselves as part of a greater whole. Integrating these perspectives into philosophical discourse fosters a sense of interconnectedness and respect for all forms of life.

Ethics and Humanity's Role in the Cosmos

Extraterrestrial messages often include calls for humanity to adopt ethical principles such as nonviolence, environmental stewardship, and compassion. These messages provide fertile ground for philosophical discussions about humanity's moral obligations, both to each other and to other life forms in the cosmos.

- **Developing a Cosmic Ethics**: If humans are not the only intelligent beings in the universe, ethical frameworks must expand to consider inter-species and interplanetary relations. Philosophers speculate on the responsibilities humans might have toward other civilizations, focusing on principles of non-interference, cooperation, and respect.
- **The Ethics of Stewardship**: Extraterrestrial messages that emphasize environmental protection align with philosophical discussions on stewardship and humanity's responsibility to protect Earth. By encouraging sustainable practices, these messages inspire humanity to act as guardians of their planet, adopting a moral responsibility that resonates on both a global and cosmic scale.

Religion and Extraterrestrial Messages: Bridging Divine and Cosmic Realities

Religion plays a pivotal role in humanity's interpretation of extraterrestrial messages, offering perspectives on creation, the divine, and the sacred. Religious narratives often include encounters with celestial beings, providing frameworks for understanding contact experiences as spiritual or divine. By integrating extraterrestrial messages into religious discourse, humanity can enrich spiritual teachings with new dimensions of cosmic awareness and interconnectedness.

Extraterrestrial Beings as Spiritual Entities

Religious traditions around the world include accounts of beings from the heavens, which are often interpreted as angels, deities, or messengers. Extraterrestrial messages may be viewed as an extension of these encounters, blending ancient spiritual teachings with modern reports of cosmic beings.

- **Angels and Divine Messengers**: In many religious texts, angelic beings serve as intermediaries between the divine and humanity, delivering messages of guidance and warning. Parallels between these figures and extraterrestrial beings encourage a reinterpretation of extraterrestrial messages as divine communications, broadening religious perspectives on the sacred.
- **Extraterrestrial Messages as Revelatory Experiences**: Some religious groups may interpret extraterrestrial messages as a form of revelation, where higher beings guide humanity toward enlightenment and spiritual evolution. This interpretation aligns with extraterrestrial messages that encourage peace, compassion, and global unity, inviting humanity to see them as a call to spiritual growth.

Integrating Extraterrestrial Concepts into Modern Faith

Modern religious leaders and scholars increasingly consider the potential for extraterrestrial life, exploring how its discovery might affect theological doctrines. Integrating extraterrestrial messages into religious teachings could transform faith traditions by incorporating universal values that transcend earthly boundaries.

- **Reinterpreting Creation Myths**: The possibility of extraterrestrial civilizations challenges traditional creation narratives, inviting reinterpretations that

accommodate the existence of life beyond Earth. By integrating extraterrestrial messages into creation myths, religions can evolve to reflect humanity's expanded understanding of the cosmos.
- **Unity and the Divine Plan**: Extraterrestrial messages that emphasize unity, compassion, and environmental stewardship resonate with the ethical principles upheld by many religious traditions. These messages can reinforce concepts of divine purpose, encouraging individuals to see unity and peace as part of a greater, divine plan that transcends human limitations.

Toward a Unified Perspective on Extraterrestrial Messages

The integration of extraterrestrial messages into science, philosophy, and religion represents a profound opportunity for humanity to expand its worldview and embrace a unified approach to understanding existence. Through scientific inquiry, humanity gains a rational framework for examining the potential reality of extraterrestrial messages, exploring concepts that challenge traditional notions of life, consciousness, and communication. Philosophy provides a space for deep ethical reflection, inviting humans to reconsider their moral responsibilities as potential members of a cosmic community. Meanwhile, religion offers pathways for spiritual integration, encouraging individuals to see extraterrestrial messages as part of a divine, interconnected plan that fosters compassion, unity, and stewardship.

As these fields converge, humanity moves closer to a holistic understanding of extraterrestrial messages, one that transcends individual belief systems and encourages a collective evolution toward peace and wisdom. By embracing this synthesis, humanity can take meaningful steps toward fulfilling the extraterrestrial call for unity, compassion, and respect for the planet—a call that echoes across scientific, philosophical, and religious domains alike.

Exploring New Frontiers: Potential Areas for Interdisciplinary Research on Extraterrestrial Messages

As humanity grapples with the complexities and implications of extraterrestrial messages, the need for interdisciplinary research becomes increasingly apparent. While individual fields such as science, philosophy, and religion offer unique perspectives on these messages, an integrated approach promises a richer, more holistic understanding. By merging scientific inquiry, philosophical thought, and religious insights, researchers can tackle the profound questions surrounding extraterrestrial contact, exploring its ethical, metaphysical, and existential dimensions. This interdisciplinary collaboration holds the potential to redefine humanity's place in the cosmos and foster a collective response that aligns with the values conveyed in these messages.

Bridging the Gaps: Why Interdisciplinary Research Is Essential

The phenomenon of extraterrestrial messaging is complex, involving elements that challenge traditional boundaries of knowledge. Interdisciplinary research allows for a convergence of expertise, enabling scientists, philosophers, theologians, and social scientists to explore these messages in a context that transcends isolated perspectives. Each discipline brings valuable insights: science offers empirical scrutiny, philosophy provides ethical frameworks, and religion adds spiritual depth. Together, they create a balanced lens through which humanity can understand and respond to the cosmic call for unity, peace, and environmental stewardship.

Scientific Inquiry Meets Philosophy and Theology: Merging Empirical and Ethical Perspectives

Astrobiology and Ethics: Understanding Extraterrestrial Life and Moral Responsibility

Astrobiology seeks to understand life beyond Earth, but its potential findings raise ethical questions that lie within the purview of philosophy and theology. Should we one day confirm extraterrestrial civilizations, humanity must consider its ethical responsibilities toward these beings, including how to interact respectfully and avoid harm.

- **Collaborative Studies on Non-Interference**: A research focus on non-interference could examine how humanity might respect the autonomy and ecosystems of other worlds. This concept, inspired by ethical principles and religious ideas of stewardship, would guide future policies on exploration and interaction with extraterrestrial beings or ecosystems, ensuring that human contact does not disturb or exploit extraterrestrial civilizations.
- **Exploring the Concept of Cosmic Kinship**: Many extraterrestrial messages emphasize unity and interconnectedness, which parallels theological and philosophical ideas of kinship with all life. Interdisciplinary research could explore how different worldviews—scientific, religious, and philosophical—approach the idea of universal kinship and the shared responsibility to protect life in all its forms.

Consciousness Studies and the Possibility of Telepathy

Some messages from extraterrestrial sources are reported to occur telepathically. This concept intersects with research on consciousness and theories in parapsychology, as well as philosophical discussions on the nature of mind and communication.

- **Examining Consciousness Beyond Biology**: In many traditions, consciousness is not confined to the human experience. Researchers from neuroscience, philosophy, and theology could collaborate to explore how consciousness might manifest in non-human, potentially extraterrestrial entities. This field might include studies on non-verbal communication, shared awareness, and the mechanics of telepathic communication, drawing on both empirical research and spiritual understandings.
- **The Ethical Implications of Telepathic Communication**: Telepathic messages, if verified, could raise concerns regarding privacy, autonomy, and consent. Interdisciplinary studies could explore ethical frameworks for respecting the boundaries of consciousness, considering how individuals and societies might protect mental privacy in a world where telepathic interactions are possible.

The Role of Sociology and Psychology: Examining the Social Impact of Extraterrestrial Messages

Sociology and psychology are essential in understanding how extraterrestrial messages affect individual and collective behavior. Interdisciplinary research can provide insights into how society absorbs and responds to these messages, whether through fear, fascination, skepticism, or spiritual awakening.

Mass Psychology and Cultural Interpretation

Cultural and psychological factors play a major role in how societies interpret extraterrestrial messages. Sociologists, anthropologists, and psychologists can study public reactions, exploring how cultural backgrounds and psychological dispositions influence perceptions and responses.

- **Analyzing Public Reaction to Apocalyptic and Environmental Messages**: Some extraterrestrial messages predict environmental crises or advocate for peace, topics that can incite strong emotional responses. Research could explore how different cultures perceive these messages, examining the role of collective consciousness and the societal factors that shape interpretations, such as media influence, historical context, and cultural values.
- **Developing Frameworks for Collective Resilience**: Messages about potential global catastrophes can create anxiety or hopelessness. Interdisciplinary research could focus on fostering resilience, developing strategies to help people process these warnings constructively. For instance, theologians might contribute spiritual coping mechanisms, psychologists might offer mental health support, and sociologists could study community-based resilience programs.

Social Movements and Ethical Action

Extraterrestrial messages that encourage peace, environmental responsibility, and unity align with themes in social justice and activism. Researchers from multiple fields can examine how these messages inspire social movements and shape ethical action, contributing to a broader understanding of collective responsibility.

- **Understanding the Role of Religion in Mobilizing Action**: Religious leaders have historically inspired

social movements, advocating for values such as compassion, justice, and peace. By analyzing how extraterrestrial messages resonate with these values, researchers can examine the potential of religious institutions to promote environmentally sustainable practices and nonviolent behavior.
- **Philosophical Perspectives on Responsibility and Solidarity**: Philosophy offers a foundation for understanding moral responsibility, particularly in the context of global challenges like environmental degradation and war. Interdisciplinary studies could explore how extraterrestrial messages reinforce these responsibilities, motivating individuals and societies to take ethical actions aligned with universal values.

Interdisciplinary Research in Technology and Communication: New Pathways for Disseminating Knowledge

With the rise of digital communication, researchers from technology, media studies, and sociology have new tools for exploring how extraterrestrial messages are spread and interpreted globally. Interdisciplinary research in this area could provide insights into how digital media shapes public perception, influences cultural narratives, and facilitates education on these profound messages.

Media Studies and the Impact of Digital Narratives

The role of media in shaping public understanding of extraterrestrial messages is significant. Interdisciplinary studies could analyze how digital narratives influence beliefs, examining both positive and negative impacts on societal attitudes.

- **Analyzing the Influence of Media on Message Interpretation**: Researchers can study the effects of media portrayals, such as documentaries, fictional works, and social media, on public beliefs about

extraterrestrial messages. By examining how these narratives align with or distort the original messages, scholars can develop strategies for promoting more accurate and constructive interpretations.
- **Developing Educational Resources and Public Outreach Programs**: To foster a balanced understanding, interdisciplinary teams could create educational materials that clarify extraterrestrial messages, presenting them in a way that is both factual and respectful. By combining insights from media studies, psychology, and sociology, these resources could encourage public engagement and critical thinking.

Technology and the Potential for Enhanced Communication

Technology, especially in fields like artificial intelligence, virtual reality, and quantum computing, offers new possibilities for understanding and disseminating extraterrestrial messages. Collaborations between technology experts, neuroscientists, and philosophers could open new avenues for exploring the mechanics of interstellar communication.

- **Virtual Reality as a Tool for Experiencing Messages**: Virtual reality (VR) technology could simulate extraterrestrial encounters, providing immersive experiences that allow individuals to explore messages in a controlled environment. These experiences, developed with input from psychologists, neuroscientists, and cultural experts, could help people better understand and process the content of extraterrestrial communications.
- **Quantum Communication and Instantaneous Transmission**: Quantum computing theories propose mechanisms for instantaneous communication across vast distances. Research into quantum communication technologies could potentially explain telepathic or

instantaneous messages reported by some individuals. By exploring this intersection of technology and communication, humanity might one day develop methods that parallel the mechanisms described in extraterrestrial messages.

Developing a New Paradigm: Ethical and Practical Guidelines for Interdisciplinary Research

As these fields converge, interdisciplinary research on extraterrestrial messages could lead to a new paradigm for ethical inquiry, exploration, and action. This paradigm would blend empirical evidence with moral philosophy and spiritual understanding, creating guidelines for humanity's response to these messages.

Establishing Codes of Conduct for Ethical Research

In interdisciplinary research, it is essential to develop codes of conduct that prioritize ethical considerations, such as respect for diverse perspectives, transparency, and integrity in the interpretation of extraterrestrial messages.

- **Respecting Cultural and Spiritual Interpretations**: As extraterrestrial messages often touch on existential questions, it is crucial to respect cultural and religious interpretations. Ethical guidelines could ensure that researchers approach these messages inclusively, valuing multiple perspectives and avoiding dogmatic interpretations.
- **Transparency and Public Involvement**: By involving the public in discussions about extraterrestrial messages, researchers can foster transparency and encourage public engagement. Public seminars, open-access research publications, and interdisciplinary forums can help build trust and promote a balanced understanding of these complex topics.

The Boundless Potential of Interdisciplinary Research on Extraterrestrial Messages

Interdisciplinary research offers humanity a unique opportunity to approach extraterrestrial messages from a holistic perspective, bridging the fields of science, philosophy, and religion to deepen our understanding. By exploring areas like astrobiology, consciousness studies, ethics, and media influence, researchers can address both the practical and metaphysical implications of extraterrestrial contact.

Through collaborative efforts, scientists, philosophers, theologians, and social scientists can illuminate the complexities of these messages, crafting a comprehensive response that resonates across all dimensions of human experience. This new paradigm not only advances knowledge but also prepares humanity to engage with the universe in ways that are respectful, responsible, and attuned to the values conveyed by our potential cosmic neighbors. As we integrate these messages into our collective consciousness, interdisciplinary research stands as a powerful tool, guiding us toward a future rooted in unity, peace, and a shared responsibility for the cosmos.

Summary of Personal and Collective Actions for Humanity's Response

This exploration of personal and collective actions inspired by extraterrestrial messages reveals a comprehensive pathway for humanity's evolution in alignment with the messages' core themes. Individually, adopting practices such as mindfulness, sustainability, and ethical lifestyle changes can help each person embody the values conveyed in these messages, fostering an inner sense of responsibility and harmony with the world. On a larger scale, humanity as a whole is urged to consider international environmental initiatives, cooperative treaties, and unified social movements aimed at peace and global unity. These collective actions provide a framework for addressing the environmental and social challenges highlighted by extraterrestrial warnings, reinforcing a shared commitment to the planet and future generations.

The integration of these messages into scientific, philosophical, and religious discourse enhances humanity's response, as each field offers unique perspectives that together create a fuller understanding. Science offers empirical insights, philosophy encourages ethical reflection, and religion provides a spiritual dimension, all contributing to a cohesive approach. The potential for interdisciplinary research amplifies this effort, allowing for a well-rounded, inclusive exploration that respects the scientific, ethical, and existential implications of extraterrestrial messages.

In embracing these personal and collective steps, humanity moves toward a future that reflects the values conveyed by these messages—a future grounded in unity, respect for life, and sustainable co-existence with the cosmos.

This comprehensive journey through the topics surrounding extraterrestrial encounters has taken us from ancient mythologies and religious apparitions to modern abduction narratives and the evolution of the Contactee Movement. We have delved into the historical context and analyzed various message themes, such as environmental warnings, nuclear disarmament, spiritual enlightenment, and apocalyptic predictions. These messages often highlight humanity's potential for growth, our collective responsibility to protect the environment, and the critical importance of peaceful co-existence.

Exploring methods of communication, we examined telepathic, symbolic, and physical interactions as conveyed through these experiences. Understanding the psychology, sociology, and cultural influences involved has further deepened our grasp on how personal and societal factors impact interpretation. From the scientific investigation into memory, perception, and telepathy to the evaluation of witness credibility and physical evidence, we have studied the rigorous methods used to assess the authenticity of these encounters.

We have also considered the intentions behind extraterrestrial messages, discussing interpretations of potential benevolence, warnings of existential risks, and even hypothetical roles in humanity's evolution. Our analysis extended to the ethical implications of extraterrestrial guidance, the possible future interactions based on these communications, and government involvement in information disclosure.

The relevance of these messages for humanity has been underscored by the pressing call for environmental stewardship, peace initiatives, and individual and collective spiritual growth. Practical responses are considered on both a personal level—such as adopting mindful, sustainable practices—and a global scale, where international treaties and social movements toward unity take center stage.

Lastly, the role of science, philosophy, and religion in interpreting and integrating these messages into human discourse points to the need for an interdisciplinary approach. By synthesizing these diverse perspectives, humanity may develop a more holistic understanding of extraterrestrial messages and their potential impact on our future. This structured exploration lays a foundation for a thoughtful, responsive approach to the profound implications of extraterrestrial encounters, empowering us to navigate our place in the universe with insight and responsibility.

Final Chapter: Embracing the Cosmic Call – Humanity's Path to Evolution

As we conclude our exploration into the messages from beyond, we find ourselves at a powerful intersection of wonder, caution, and responsibility. These messages—whether conveyed through ancient myths, religious visions, modern abductions, or cryptic symbols—are more than isolated, mysterious events. They reflect a consistent call from the cosmos, a call that transcends time, culture, and belief, urging humanity to look beyond our immediate experiences toward a more profound understanding of our place in the universe. These encounters, whether literal or metaphoric, invite us to consider that the broader vision for humanity's evolution may hinge on our ability to heed and interpret these messages.

The Messages in Context: A Cosmic Perspective on Human Responsibility

At the heart of these extraterrestrial encounters lies an urgent theme: a responsibility to our environment, our species, and our shared destiny. The environmental warnings, often articulated through abduction experiences and spiritual visions, highlight a critical truth: our planet's well-being is inextricably linked to our own. These messages encourage us to confront and reverse the damage we've inflicted upon Earth, to view nature not as a resource but as a living entity deserving of respect and stewardship. They remind us that safeguarding the environment is not simply a survival imperative but a step toward embracing our role as cosmic stewards, responsible for the harmonious existence of life across dimensions.

Evolution Through Peace: The Extraterrestrial Appeal for Disarmament

Equally prominent in these messages is the insistence on peace and disarmament, particularly regarding nuclear weapons. From the era of the Contactee Movement in the Cold War period to recent accounts by military personnel, warnings against nuclear destruction suggest that our choices on Earth ripple far beyond our planet. Extraterrestrials, often portrayed as advanced beings with technologies beyond our grasp, seem deeply concerned with humanity's potential to self-destruct. In their appeal for peace, we find a model of a society that has transcended its violent impulses. For humanity, disarmament represents not just a moral victory but an evolutionary leap toward a future where cooperation, unity, and trust replace aggression, division, and fear.

Spiritual and Consciousness Evolution: Awakening the Higher Self

Many messages encourage a path of spiritual enlightenment, urging humanity to transcend materialism and foster inner growth. These calls for spiritual evolution resonate across encounters, suggesting that the true measure of progress is not in technology but in consciousness. By embracing practices like meditation, mindfulness, and compassion, we align ourselves with the transformative energies reflected in these messages. This evolution of consciousness is an invitation to understand reality beyond the physical, to tap into our latent potential, and to cultivate a sense of unity that spans individuals, nations, and even species. In essence, the cosmic call for spiritual growth points to a collective awakening, where humanity realizes that we are not isolated beings but interconnected souls within a vast, intelligent universe.

Unity in Diversity: A Global and Cosmic Citizenship

The commonalities in extraterrestrial messages serve as a reminder that humanity's diversity is a strength. While different cultures, religions, and philosophies have interpreted these messages in unique ways, the essence remains remarkably consistent: we are all part of a larger, interconnected system. These encounters encourage us to transcend boundaries and divisions, to cultivate unity not just among ourselves but with the greater cosmos. They suggest a form of cosmic citizenship in which humanity is a member of a universal community, bearing mutual responsibilities, and bound by shared principles of respect, harmony, and growth.

Humanity's Path Forward: Embracing Our Cosmic Potential

So where do we go from here? The cosmic messages, with their spiritual and practical guidance, provide a framework for personal and collective transformation. Individually, we are called to adopt practices of inner peace, environmental stewardship, and compassionate action. As a global community, we are challenged to build a society that reflects our highest ideals—one that is capable of coexisting peacefully, protecting our planet, and exploring the mysteries of the universe with humility and curiosity.

But perhaps the most profound insight from these encounters is the realization that humanity's evolution is not simply a biological process but a conscious choice. As we stand at the threshold of a new era, we are invited to decide what kind of species we wish to become. Will we continue on a path of division, isolation, and environmental neglect, or will we embrace the vision offered by these messages—one of unity, responsibility, and cosmic harmony?

A New Paradigm: Humanity's Role as Cosmic Stewards

In accepting these messages, we acknowledge that our destiny is not confined to Earth but is intricately woven into the fabric of the cosmos. Our actions, thoughts, and intentions have far-reaching implications, impacting not only ourselves but also the broader universe. To embody the lessons imparted through these encounters is to step into our role as cosmic stewards, guardians of not just our own world but of the harmony and balance within a vast, interconnected reality. This paradigm shift may seem daunting, but it represents a profound leap toward a future where humanity thrives in alignment with universal laws and cosmic intelligence.

The Closing Message: An Invitation to Transform

As we conclude this journey, we recognize that these messages are more than warnings or prophecies; they are invitations to transform. They remind us of our potential to evolve, to contribute positively to the universe, and to understand our place within a greater cosmic order. We are at a pivotal moment in history, with the opportunity to take these messages to heart and to create a world that reflects the values of peace, unity, and respect for all life.

In embracing this call, humanity embarks on a path of conscious evolution, grounded in compassion, wisdom, and a deep sense of purpose. This is the legacy we are invited to build—a legacy that honors not only those who have come before us but also those who will follow. The messages from beyond may originate from realms unknown, but their purpose is clear: to guide us toward a future where humanity realizes its highest potential as a responsible, enlightened member of the cosmic family. Let us walk this path with courage, integrity, and an open heart, ready to become the stewards of a world—and a universe—that we are just beginning to understand.

SOURCES

1. Historical Context of Extraterrestrial Encounters and Messages

- **Ancient Encounters and Mythologies**
    - Daniken, Erich von. *Chariots of the Gods? Unsolved Mysteries of the Past.* Putnam, 1968. (One of the earliest modern works suggesting ancient texts describe extraterrestrial contact.)
    - Narayan, R. K. *The Mahabharata: A Shortened Modern Prose Version of the Indian Epic.* University of Chicago Press, 1978. (Translation and interpretation of ancient Indian epics with descriptions of flying vehicles and cosmic wars.)
    - Jung, Carl G. *Flying Saucers: A Modern Myth of Things Seen in the Skies.* Princeton University Press, 1978. (Psychological exploration of UFO sightings and symbolism in modern and ancient times.)
- **Religious Apparitions with Potential UFO Elements**
    - Zimdars-Swartz, Sandra L. *Encountering Mary: From La Salette to Medjugorje.* Princeton University Press, 1991. (Detailed study of Marian apparitions, including Fatima, with discussion of possible UFO interpretations.)
    - Vallee, Jacques, and Chris Aubeck. *Wonders in the Sky: Unexplained Aerial Objects from Antiquity to Modern Times.* TarcherPerigee, 2010. (Chronicles documented sightings and otherworldly encounters throughout history.)
- **The Contactee Movement (1950s–1970s)**
    - Peebles, Curtis. *Watch the Skies! A Chronicle of the Flying Saucer Myth.* Smithsonian Institution Press, 1994. (Thorough history of the UFO phenomenon, including the contactee movement.)

- - Adamski, George. *Inside the Space Ships.* Abelard-Schuman, 1955. (One of the original accounts by a contactee claiming direct encounters and messages from extraterrestrials.)
  - **Modern Abduction Phenomenon and Message Themes (1970s–1990s)**
    - Mack, John E. *Abduction: Human Encounters with Aliens.* Scribner, 1994. (Psychiatrist's perspective on abduction cases and the messages reported by abductees.)
    - Hopkins, Budd. *Missing Time: A Documented Study of UFO Abductions.* Ballantine Books, 1988. (Detailed account of abduction cases and themes in the messages received.)

2. Analyzing Types of Messages Received in Encounters

- **Environmental Destruction Warnings**
  - Druffel, Ann. *Firestorm: Dr. James E. McDonald's Fight for UFO Science.* Wild Flower Press, 2003. (Covers environmental and social concerns within UFO messages.)
  - Good, Timothy. *Need to Know: UFOs, the Military, and Intelligence.* Pegasus Books, 2007. (Discusses instances of UFO messages addressing humanity's environmental and social issues.)
- **Nuclear Weapons and War Warnings**
  - Jacobs, David M. *The Threat: Revealing the Secret Alien Agenda.* Simon & Schuster, 1998. (Explores themes of nuclear threat in abduction messages.)
  - Hastings, Robert. *UFOs and Nukes: Extraordinary Encounters at Nuclear Weapons Sites.* AuthorHouse, 2008. (Extensive

documentation of UFO sightings near nuclear facilities and associated warnings.)
- **Spiritual Enlightenment and Evolution**
    - Streiber, Whitley. *Communion: A True Story.* William Morrow, 1987. (Explores the spiritual messages and personal evolution experienced in abduction encounters.)
    - Mack, John E. *Passport to the Cosmos: Human Transformation and Alien Encounters.* Crown, 1999. (Documents how abductees' experiences often carry themes of spiritual growth.)

3. Methods of Communication in Extraterrestrial Encounters

- **Telepathic Communication**
    - Rodeghier, Mark. "Telepathic Communication during Alien Abductions: The Role of Hypnosis in Recalling Encounters." *Journal of Scientific Exploration*, vol. 16, no. 2, 2002, pp. 229–245.
    - Mack, John E. *Abduction: Human Encounters with Aliens.* Scribner, 1994. (Covers telepathy as a common method of communication in abductions.)
- **Symbolic and Visual Communication**
    - Jung, Carl G. *Flying Saucers: A Modern Myth of Things Seen in the Skies.* Princeton University Press, 1978. (Analyzes symbols in UFO experiences and their cultural meaning.)
    - Glaser, Hans. *The 1561 Nuremberg Celestial Event.* (A woodcut illustrating a UFO-like phenomenon; discussed in *Wonders in the Sky* by Jacques Vallee and Chris Aubeck, 2010.)
- **Physical Contact and Communication through Abductions**
    - Hopkins, Budd. *Intruders: The Incredible Visitations at Copley Woods.* Random House,

1987. (Examines abduction cases involving physical contact and message exchange.)
- Streiber, Whitley. *The Secret School: Preparation for Contact*. HarperOne, 1997. (Explores physical and sensory experiences in alien encounters.)

4. Psychological, Sociological, and Cultural Perspectives

- **The Psychology of Encounters**
    - Appelle, Stuart. "The Abduction Experience: A Critical Evaluation of Theory and Evidence." *The Journal of American Folklore*, vol. 109, no. 432, 1996, pp. 106–118.
    - Mack, John E. *Passport to the Cosmos*. Crown, 1999. (Analyzes the psychological impact of encounters.)
- **Mass Hysteria and Social Influence**
    - Sheaffer, Robert. *UFO Sightings: The Evidence*. Prometheus Books, 1998. (Examines cases of mass hysteria related to UFO sightings.)
    - Bartholomew, Robert E., and Hilary Evans. *Outbreak!: The Encyclopedia of Extraordinary Social Behavior*. Anomalist Books, 2009. (Covers the social influence and hysteria linked to UFO sightings.)
- **Cultural and Religious Influences on Interpretation**
    - Bullard, Thomas E. *The Myth and Mystery of UFOs*. University Press of Kansas, 2010. (Looks at cultural and religious influences on UFO interpretations.)
    - Lewis, James R., and Glenn W. Shuck. *The Gods Have Landed: New Religions from Other Worlds*. SUNY Press, 1995. (Analyzes extraterrestrial messages in the context of spiritual movements.)

5. Scientific Investigation of Message Encounters

- **Role of Hypnosis in Message Retrieval**
    - Spanos, Nicholas P., and Jack Gottlieb. "Hypnosis and UFO Abduction." *The Journal of Abnormal Psychology*, vol. 104, no. 4, 1995, pp. 603–611.
    - Klass, Philip J. *UFOs Explained*. Random House, 1974. (Critical evaluation of hypnosis in abduction cases.)
- **Neurological Studies on Perception and Memory**
    - Mack, John E. *Abduction: Human Encounters with Aliens*. Scribner, 1994. (Examines neurological impacts of encounters.)
    - Persinger, Michael A., and Gyslaine F. Lafreniere. *Space-Time Transients and Unusual Events*. Nelson-Hall, 1977. (Discusses neurological models of unusual experiences.)
- **Studies on Telepathy and Consciousness**
    - Radin, Dean. *The Conscious Universe: The Scientific Truth of Psychic Phenomena*. HarperOne, 1997. (Explores telepathy research and consciousness.)
    - Tart, Charles T. *The End of Materialism: How Evidence of the Paranormal Is Bringing Science and Spirit Together*. New Harbinger, 2009. (Discusses consciousness studies related to telepathic communication.)

---

6. Patterns and Commonalities in Extraterrestrial Messages

- **Identifying Common Themes**
    - Appelle, Stuart. "The Abduction Experience: A Critical Evaluation of Theory and Evidence." *The Journal of American Folklore*, vol. 109, no. 432, 1996, pp. 106–118.

- Druffel, Ann. *Firestorm: Dr. James E. McDonald's Fight for UFO Science.* Wild Flower Press, 2003. (Highlights recurring themes in UFO messages.)
- **Comparative Analysis with Other Phenomena**
    - Ring, Kenneth. *The Omega Project: Near-Death Experiences, UFO Encounters, and Mind at Large.* William Morrow, 1992. (Compares UFO messages with near-death experiences.)
    - Vallee, Jacques. *Dimensions: A Casebook of Alien Contact.* Ballantine Books, 1988. (Draws parallels with shamanic and mystical experiences.)

7. Interpretations and Theories about Extraterrestrial Intentions

- **Benevolent vs. Malevolent Intent**
    - Jacobs, David M. *The Threat: Revealing the Secret Alien Agenda.* Simon & Schuster, 1998.
    - Vallée, Jacques. *Messengers of Deception: UFO Contacts and Cults.* And/Or Press, 1979.
- **Extraterrestrial Civilizations' Perspective on Humanity**
    - Däniken, Erich von. *Chariots of the Gods? Unsolved Mysteries of the Past.* Putnam, 1968.
    - Streiber, Whitley. *The Key: A True Encounter.* Walker & Company, 2001.

8. Evaluating the Credibility and Authenticity of Messages

- **Assessing Witness Credibility**
    - Hynek, J. Allen. *The UFO Experience: A Scientific Inquiry.* Henry Regnery, 1972.

- Bullard, Thomas E. *UFO Abductions: The Measure of a Mystery.* Volume 1, BFI Publishing, 1987.

9. The Message's Relevance for Humanity and Our Future

- **Environmental Warnings and Human Responsibility**
    - McKenna, Terence. *The Archaic Revival.* HarperCollins, 1991.
    - Fowler, Raymond. "The Human Potential Movement and Alien Contact." *Journal of Humanistic Psychology*, vol. 37, no. 3, 1997.

10. Ancient and Pre-Modern Cases of Potential Extraterrestrial Messages

- **Earliest Accounts of Otherworldly Messages**
    - Vallee, Jacques, and Chris Aubeck. *Wonders in the Sky.* TarcherPerigee, 2010.
    - Ezekiel's Vision in *The Bible* (Book of Ezekiel).
- **Roman and Greek Accounts of Celestial Phenomena**
    - Tacitus. *Annales.*
    - Plutarch. *Lives.*
- **Medieval and Renaissance Accounts of UFO-like Events**
    - Glaser, Hans. *The 1561 Nuremberg Celestial Event* woodcut.

www.ingramcontent.com/pod-product-compliance
Lightning Source LLC
Chambersburg PA
CBHW051047230426
43666CB00012B/2596